The Banchine Wasps

(Ichneumonidae: Banchinae)

of the British Isles

J. P. Brock

Horniman Museum

Figures by

J. P. Brock

Colour plates by

O. Retka & A. Polaszek

Natural History Museum

J. P. Brock

Published for the   Royal Entomological Society
The Mansion House
Bonehill
Chiswell Green Lane
Chiswell Green
St Albans
AL2 3NS
www.royensoc.co.uk

By the   Field Studies Council
Unit C1
Stafford Park 15
Telford
TF3 3BB
www.field-studies-council.org

FSC

BRINGING
ENVIRONMENTAL
UNDERSTANDING TO ALL

ISBN: 978 1 910159 01 9

Contents

Abstract	iv
Acknowledgements	v
Introduction	1
Historical	1
Taxonomy	1
Glossary	2
Identifying the subfamily Banchinae	7
Identification procedure	10
Nomenclature	11
Biology	12
Distribution, abundance and phenology	14
Collecting	15
Checklist	16
Keys to tribes of Banchinae	25
Tribe Glyptini	26
Keys to the genera/subgenera of Glyptini	26
Telutaea	28
Apophua	28
Diblastomorpha	31
Conoblasta	32
Glypta (sections and species groups)	38
nigrina section	40
consimilis section	40
glypta section	41
resinanae group	41
haesitator group	42
bifoveolata group	47
scutellaris group	48
mensurator group	49
pedata group	50

Tribe Atrophini	53
Keys to the genera of Atrophini	54
Syzeuctus	55
Arenetra	56
Alloplasta	57
Lissonota (subgenera)	58
Meniscus	60
Lampronota	60
Lissonota s. str.	68
Loxonota	72
Campocineta (species groups)	75
admontensis group	78
linearis group	79
coracina group	80
clypealis group	82
saturator group	83
gracilenta group	90
versicolor group	92
dubia group	94
buccator group	94
Cryptopimpla	108
Tribe Banchini	113
Keys to the genera of Banchini	114
Exetastes	114
Banchus	121
Rynchobanchus	126
Appendix (Hosts and parasitoids)	127
References and further reading	132
Index	135
Colour plates	141

Abstract

This book covers the 138 British species in the subfamily Banchinae. It is based on a full taxonomic revision, including reference to the accessible type specimens. The classificatory procedure used here is predominantly 'utilitarian' in nature, since the primary objective is to render identification of banchine ichneumonids as simple as possible. Keys are given to all species – and those for *Glypta*, *Lissonota*, *Cryptopimpla and Exetastes* incorporate a full revisionary treatment. The creation of species groups has been necessary in the two very large genera *Glypta* and *Lissonota* – in particular, within the large subgenus *Campocineta* in *Lissonota*. A general introduction, species accounts and an account of all verified host relationship are also provided. Ten new species are described.

Acknowledgements

I would like to express my gratitude to the many individuals and institutions that have contributed to this project. Special thanks to: the Horniman Museum; Mark Shaw, for the loan of a very large body of material resulting from long-term Malaise-trapping and host-rearing; Gavin Broad, David Notton and Mike Fitton for facilitation of access to the Natural History Museum collection in London. Tony Irwin provided an opportunity to study the Bridgman types at Norwich Castle Museum. Bert Viklund of the Swedish Museum of Natural History in Lund arranged facilities for me to examine the Thomson collection. Robert Nash arranged access to Ulster Museum material. George McGavin provided access to the Hope Department of Entomology material in the Oxford University Museum. Additional information and specimens relating to material collected in the UK came from Bill Ely – data from Yorkshire and other northern English museums, and from his own field work. Phil Sterling donated reared material of *Glypta ulbrichti*. The late A.W. Jones loaned material from his private collection. Ian McLean organised a British Entomological Society workshop for testing earlier drafts of keys. The late Klaus Horstmann afforded valuable discussion regarding the taxonomy of Banchinae – with special reference to the location of 'lost' type material. Kees Zwakhals – for help in obtaining the cover photograph.

The following institutions allowed me to borrow type (and other) material from their collections:

Evolutionsmuseet, Zoologi, Uppsala
Finnish Museum of Natural History, Helsinki
Horniman Museum, London
Hungarian Natural History Museum, Budapest
Institut für Zoologie, Salzburg
Linz Museum, Austria
Manchester University Museum
Musée de Zoologie, Lausanne
Nationaal Natuurhistorisch Museum, Netherlands
National Museum of Scotland, Edinburgh
Natural History Museum, London
Naturhistorisches Museum, Admont
Norwich Castle Museum, Norwich
Oxford University Museum, Oxford
Staatliches Museum für Naturkunde, Stuttgart
Swedish Museum of Natural History, Stockholm
Theodor-Boveri-Institut fur Biowissenschaften, Würzburg
Ulster Museum, Belfast
Universitets Zoologisk Museum, Copenhagen
Universytet Wroclawski, Wroclaw, Poland
Zoologische Museum, Berlin
Zoologische Staat Museum, Munich

The following abbreviations are used throughout the present document with respect to institutions and individuals in the UK:

HM: Horniman Museum, London
JPB: J. P. Brock
MU: Manchester University Museum

NCM: Norwich Castle Museum
NHM: Natural History Museum, London
NMS: National Museum of Scotland, Edinburgh
WE: W. Ely

The following individuals arranged postal access to essential type and other material from the European continent: Kees van Achterberg, Milan Chvala, Sándor Csősz, László Forró, J. Götze, Klaus Horstmann, F. Koch, Karl-Heinz Krisch, Hans Mejlon, Till Osten, Martin Schwarz, Hege Vårdal, and Marek Wanat.

Facilities for dealing with loans between 2007 and 2010 were provided through Jenny Clack and W. Foster, Cambridge Museum of Natural History.

Management staff of National Nature Reserves and Sites of Special Scientific Interest authorised collecting in many important sites, notably: Ashtead Common, Surrey; Bure Marshes, Norfolk; Chippenham Fen, Cambridgeshire; Chobham Common, Surrey; Flanders Moss, Stirlingshire; Folkestone Warren, Kent; Stodmarsh, Norfolk; Sydenham Wood, London; Thursley Common, Surrey; Wicken Fen, Cambridgeshire; Windsor Forest, Surrey; Winterton, Norfolk.

Most significantly, the greater part of research for the present publication was facilitated through the generosity of the Directors and Trustees of the Horniman Museum, Forest Hill, London.

Introduction

Historical

The first comprehensive monograph of European Ichneumonidae was written by J.L.C. Gravenhorst in 1829, and in this work we find the foundation for all subsequent research on the family. Two other early publications of special relevance to the Banchinae were those authored by A.E. Holmgren (1860) and C. G. Thomson (1877, 1889a, b). Between them, Gravenhorst, Holmgren and Thomson described over 40% of the existing British Banchinae species, the remainder being shared amongst 26 other authors.

In Great Britain, the first listing of native Ichneumonidae was carried out by Thomas Desvignes (1856). Closer to the 20th century, J.B. Bridgman (1882-1890) made several important contributions to our knowledge of the indigenous Banchinae, including descriptions of a number of new species. At a later date, Claude Morley wrote a five volume monograph covering the entire British Ichneumonidae, the Banchinae being included in Vol. 3 (1908). Many species can still be named using Morley's work, given access to a recent checklist in order to trace changes in nomenclature. However, the absence of very many Banchinae now known to occur in the UK obviously creates practical difficulty here – and the lack of any realistic assessment of infraspecific variation by Morley greatly exacerbates the problem. In addition, Morley's accounts of distribution, abundance, and host preferences were greatly marred by misidentification of both parasitoids and hosts.

Further into the twentieth century, J.F. Perkins of the NHM examined a large proportion of the type specimens of British Ichneumonidae, although the greater part of his work was never published. More recently, H.T. and M. Townes carried out an advanced study of the North American Banchinae (1978), and J.F. Aubert created an extensive catalogue of the European species (1978). Aubert also provided keys to *Glypta* and *Lissonota* that were to a certain extent 'provisional' in nature. However, bearing in mind the fact that the latter component of Aubert's work constituted an appendix to what was intended primarily as a catalogue, we do actually encounter a highly significant improvement over earlier efforts. Both Townes and Aubert made valuable cross-reference to Perkins' unpublished work, and the Townes study also added many novel characters to the classical repertoire. In addition to the publications already cited, Kuslitsky published much valuable original work on the tribe Glyptini (summarised in Kasparyan, ed., 1981), and the genus *Banchus* was covered by Fitton (1985) for the Western Palaearctic region. More recently, the subgenus *Loxonota* (*Lissonota*) was revised by Rey de Castillo (1992).

Taxonomy

Banchines are a subdivision of the old 'Pimplinae', which was originally defined as a group having long ovipositor plus sessile abdomen (metasoma). Two small subfamilies have been separated from Banchinae in relatively recent times: Stilbopinae and Neorhacodinae, both of which can be excluded on the basis of information given below.

The genera of Banchinae are mostly clearly defined and easily identified – a situation which is almost unique amongst the larger subfamilies of Ichneumonidae. In addition, the most species-rich genus *Lissonota* can be conveniently divided into a number of species groups. At the other end of the spectrum, species definition is unusually difficult in the *Glypta* subgenus *Conoblasta*.

Comparing the British banchine fauna with that of (the unrelated) Pimplinae (*sensu* Townes, 1969), we find the latter taxon carrying 36 genera, of which the three largest (*Scambus, Dolichomitus, Pimpla*) account for around one tenth of the total species count. By way of stark contrast, the somewhat more species-rich Banchinae contains only 11 genera, of which the two largest (*Glypta* and *Lissonota*) account for over 70% of the entire subfamily.

So far as the present study is concerned, there are perhaps four or five undetermined *Lissonota* species in collections – in all probability, at least the same number yet to be encountered. On this basis, the genus in question may ultimately be found to hold over 70 British species. Given that the same principle applies to *Glypta*, we might expect a final total of around 160 banchine species occurring in the British Islands, once more extensive collecting and rearing has been carried out.

Glossary

There follows an account of characters used in determining Banchinae. This includes reference to some terms relating to the head that are used differently from the customary view, thus it is essential to become familiar with these definitions at the outset. In addition, two new terms relating to the mesopleurum and sternum are added to the usual repertoire.

Head characters

1. **Maximum temple length:** this dimension is measured between occipital carina and compound eye, at a position no further forward than that point at which the orbit of the eye turns forwards (usually marked by a slight angulation) (Fig. 1a - **MaxTL**).

2. **Minimum genal width:** measured between occipital carina and compound eye, no further ventrad than that point at which the orbit turns forward (thus no longer running parallel to the occipital carina) (Fig. 1b - **MinGL**).

3. **Separating gena from temple:** the temple and gena form a continuous surface in very many ichneumonids. However, in some banchines, they are non-coplanar, such that their intersection is marked by a disjunction. In actual practice, the lack of any universally recognisable definition between gena and temple does not affect usage of the above dimensions based on maximum temple length/minimum genal width. While comparative characters derived from gena and temple can be difficult to pinpoint to a precise location, this fact is more than offset by the normal range of variation to which they are subject.

4. **Length of compound eye:** this dimension is equivalent to the maximum transverse diameter of the compound eye.

5. **Malar space:** this trait has been used differently by different authors. For the present study, the measurement is taken between the top edge of the mandible base and the nearest point on the compound eye. This is often deemed necessary, since it is common for the line of the mandible base to lie at a fairly steep angle relative to the lower contour of the compound eye (Fig. 1b).

6. **Antenna** (Fig. 1b): The large basal segment of the antenna is the scape, which occasionally offers structural characters in Banchinae. The scape is followed by the *pedicellus* and *annellus* – two reduced segments, neither of which is of taxonomic interest in the present context. The main part of the antenna is the *flagellum* – of which the first segment probably offers the best

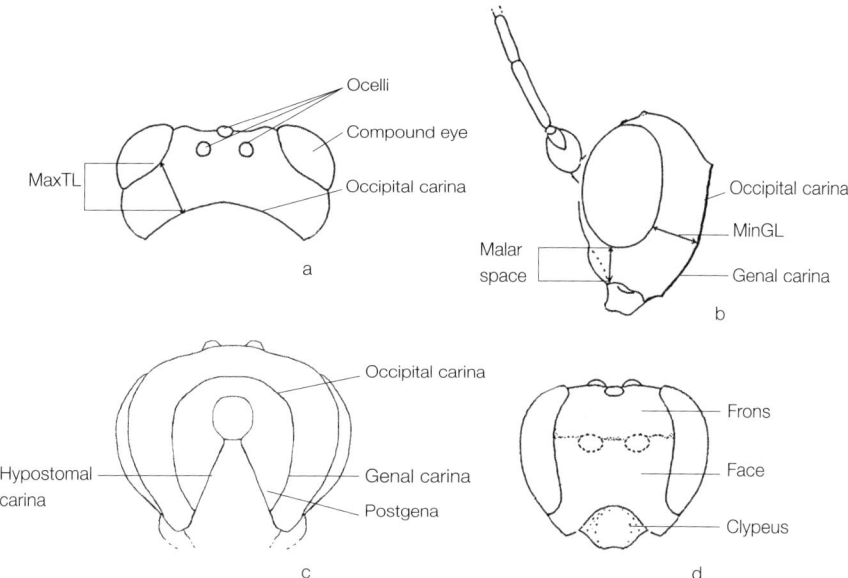

Figure 1. Head. (a) Dorsal view; (b) Lateral view; (c) Posterior view; (d) Frontal view.

source of diagnostic traits in the Ichneumonidae in general. Here, we are usually concerned with the length:breadth ratio – in which instance 'breadth' is equivalent to the maximum (usually the apical) dimension. The more distal flagellar segments occasionally offer additional taxonomic characters – again usually in the form of length:breadth ratios, although sometimes as divergent structural traits, e.g. as in the moniliform (bead-like) state of most *Cryptopimpla* species. The basal flagellar segments may sometimes be provided with a few enlarged sensilla that are of some taxonomic interest. I have omitted reference to these structures, for the simple reason that they are of a 'deciduous' nature – thus difficult to incorporate into practical identification keys.

7. **Ocelli and compound eyes** (Fig. 1a): The dimensions of the interspace between the posterior ocelli as this relates (a) to the neighbouring compound eyes, and (b) to the occipital carina behind, are of fundamental importance in species discrimination in Banchinae (as indeed, in a majority of parasitoid Hymenoptera).

8. **Occipital carina** (Fig. 1b, c): Apart from its relationship with the hind ocelli, the occipital carina may itself exhibit useful diagnostic traits, e.g. in being interrupted or angled centrally.

9. **Face and frons** (Fig. 1d): Face and frons may be quite differently sculptured from one another, and the manifestation of sculpturing may differ from one taxon to another. Aspects of colour pattern frequently manifest themselves in these regions (while at the same time being subject to a large degree of infra-specific variation – as indeed, are colour characters in general).

10. **Clypeus** (Fig. 1d): The form of the apical margin of the clypeus affords good taxonomic characters at and around level of tribe in Banchinae (also within the tribe *Banchini*). In particular, there may be angular clypeal margins – in contrast to the usual evenly arcuate apical rim. In addition, the usual discrimination from the face may sometimes be more or less effaced.

11. **Mandibles**: The mandibles of Banchinae are quite uniform in general shape – with the exception of *Banchus* and its near relatives. Nevertheless, there may be some degree of variation in sculpture from one species to another, e.g. in the degree to which punctation is in evidence.

Thoracic characters

1. The **notauli** are a pair of linear impressions arising from the front edge of the mesonotum dorsally and continuing for some distance behind (Fig. 2a). They are usually only weakly developed in Banchinae.

2. The **speculum** (the normally unsculptured area lying towards the apical margin of the mesopleurum) (Fig. 2b) is said to be '*open*' when it continues to the hind edge of the mesopleurum – and '*closed*', when interrupted by a zone of punctation. Comparative measurements relating to the mesopleural speculum are taken as follow:
   (i) distance between the posterior mesopleural suture and anterior margin of the speculum – as a fraction of (ii) distance between mesopleural suture and epicnemium. This index is taken along a horizontal line passing through the speculum at its widest part. (See Fig. 2b).

   In the diagnostic keys, the above index is abbreviated as 'speculum reaching [fraction] distance between mesopleural suture and epicnemium'.

3. The **mesosulcus** (Fig. 2c): a groove running along the mid line of the mesothoracic sternum – is said to be '*open*' when it continues beyond the apex of the mesosternum, '*closed*' when it is interrupted at that point by one or more strong transverse carinae.

4. **Wings:** there are three aspects of special interest here: (i) the form of the **areolet** (cell 2Rs) in the fore wing (Fig. 2d, e), which may be present or absent – and which also may take different

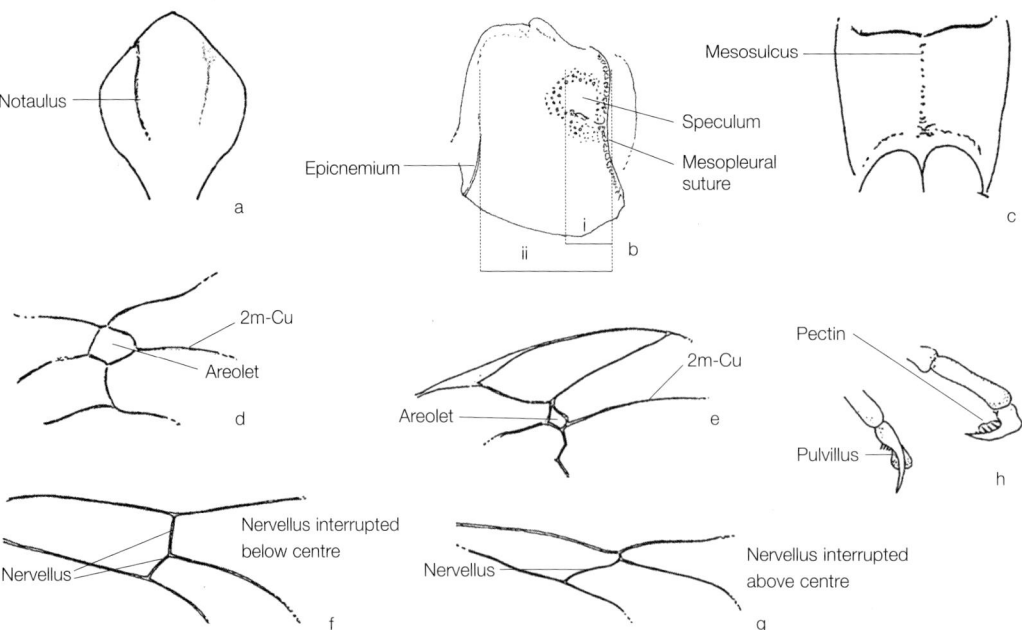

Figure 2. Mesonotum. (a) Dorsal view, notaulus. (b) Lateral view.
Mesosternum (c) mesosulcus. Fore wing (d & e) areolet. Hind wing, nervellus, (f) broken below centre, (g) broken above centre
Claws (h) Pulvillus and pectin.

4

contours when present; (ii) the position on the **areolet** at which the second medio-cubital vein (2m-Cu) joins; (iii) the **nervellus** in the hind wing – a compound vein made up of the first section of cubitus plus the cubito-anal vein (Cu / Cu-a), which may be 'intercepted' above or below the middle by the Cu-1 vein (Fig. 2f, g).

5. **Legs:** colour pattern is a commonly used trait of the legs – especially of the hind pair. In addition, important points of discrimination are to be found on the tarsal claws. In the first place, the relative length of the claws and the **pulvillus** (a pad-like structure between the claws) (Fig. 2h) is of prime significance in subdividing the very large genus *Lissonota*. Secondly, the extent to which the claw itself may be 'pectinate' (i.e. provided with a comb of strongly sclerotised teeth) is a further important higher group trait within the same genus.

Abdominal characters

1. **Propodeum** (Fig. 3a & b): In purely functional terms, the propodeum is integrated with the thorax. This is a peculiarly hymenopteran feature (although paralleled by an aposomatic South American genus of syntomid moths). Morphologically speaking, the propodeum is in fact of abdominal origin. The 'areation' of the propodeum (subdivision of its surface via a pattern of raised carinae) offers many taxonomic traits that are very widely used throughout the Ichneumonidae. In Banchinae, there appears to be an unusual degree of infraspecific variation with regard to areation of the propodeum in some species groups. Nevertheless, useful characters are to be found here in many taxa.

2. **Metasoma**: This is the 'functional abdomen' of ichneumonids (as in a majority of parasitoid wasps). A great many valuable characters (see below) are to be found here, both in structure and in colour pattern:

   Petiolar segment: The first metasomal segment is generally differentiated from the remainder by virtue of being narrowed towards base, and 'petiolate' to a greater or lesser extent (i.e. stalked, see Fig. 3, d). Tergite 1 is usually sculptured quite differently from the remainder of the

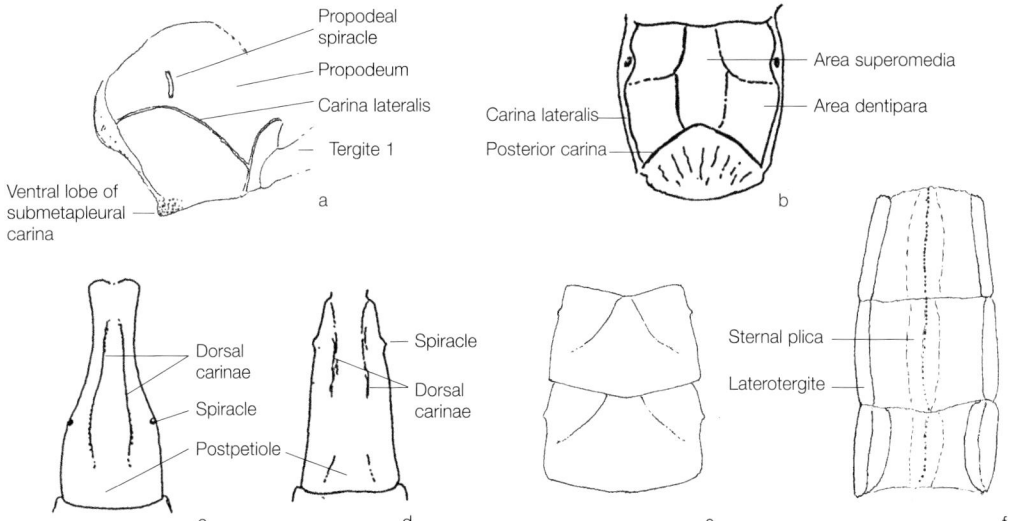

Figure 3. Propodeum. (a) Lateral view; (b) Dorsal view.
Metasoma: tergite 1 (c) 'petiolate'; (d) 'sessile'. (e) Glyptini. Metasoma, dorsal view (f) laterotergites.

metasoma, and characters drawn from this very often afford useful taxonomic traits. In particular, the width:length ratio (width taken across postpetiole, i.e. its widest point), is often of crucial importance in discrimination of species, as well as of supra-specific categories). In addition, the form of the dorsal carinae of tergite 1 and sculpturing of the postpetiole (apical sector) (Fig. 3c, d) is frequently of importance.

Central tergites: The first two metasomal tergites beyond the petiolar segment (usually called the 'central' tergites) generally afford inter-specific structural traits of the highest value. These usually take the form of surface sculpture, and / or the presence of impressed areas running diagonally or transversely across one or more of the tergites (as Fig. 3e) – also variation in the length:width ratio. In addition, colour pattern often provides useful (albeit rather variable) characters for species discrimination.

Apical tergites: Of special significance here is the degree to which the tergal pubescence extends onto the dorsum of the pre-apical tergites – an important character in some *Lissonota* species groups.

Laterotergites: These structures form lateral flaps on the metasomal tergites, and generally either hang downwards or are folded more or less horizontally beneath ('*inflexed*') (Fig. 3f). Two aspects of the laterotergites may be of special interest: (i) their length: breadth ratio; (ii) the density of pubescence on the laterotergites of metasomal segments 3 and 4. As with the form of the apical tergites (see above), these traits are often obscured due to *post mortem* contraction. For this reason, I have generally kept them in low profile in this Handbook.

Ovipositor: Ovipositor length is a fundamental diagnostic character in many higher groups within Ichneumonidae. However, this trait does exhibit a much greater degree of variation in Banchinae than we are often led to believe. The shape of the ovipositor may also offer divergent features, such as the degree of curvature, or in the extent to which the width changes between base and apex. Unlike many other ichneumonids, the apex of the ovipositor exhibits very little variation between species, e.g. being quite bereft of the usual dentition observed in the 'true' Pimplinae. Most important of all is the fact that for *ovipositor length*, it is the ovipositor sheaths that should be measured rather than the ovipositor. The ovipositor is prone to much size variation, due to post-mortem effects.

Identifying the subfamily Banchinae

Identifying the subfamily Banchinae is not a particularly difficult task – at least for the female sex, where the ovipositor usually forms a good landmark. The following key provides a much simplified guide, which should serve for a majority of specimens encountered. It does *not* attempt to form a substitute for a fully exhaustive key to subfamilies of Ichneumonidae – and it will work best with females (where ovipositor structure can be of prime importance).

1. Tergite 1 of metasoma petiolate, i.e. distinctly narrowing in basal part; usually curved at juncture with postpetiole, and with the spiracles situated beyond middle of segment (Fig. 1); fore wing with areolet (cell 2Rs) frequently pentagonal .... [*other subfamilies*]

–   Tergite 1 sessile, i.e. not or only weakly narrowing in basal section; at most weakly bent at position of postpetiole, the spiracles situated before middle of segment (Fig. 2) (areolet 4-sided or absent (Figs 3, 4)) 2

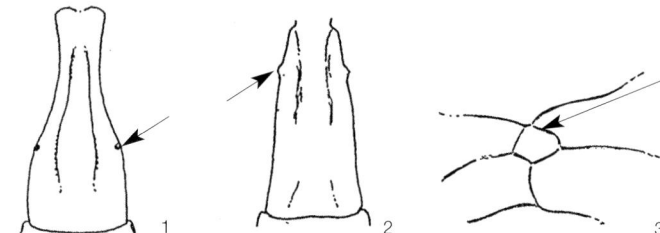

2. Central tergites of metasoma with an inverted V-shaped impression, and with no transverse groove (Fig. 5) .... **Banchinae: Glyptini** (p. 26)

–   Tergites lacking such sculpturing (excepting rarely, when the impressions are joined by a horizontal groove before posterior margin of tergite) 3

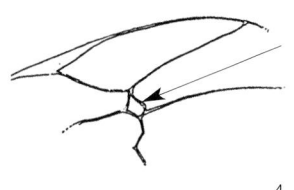

3. Areolet large and rhomboidal (Fig. 3); clypeus discriminated from face (Fig. 6); large species – wing length over 7 mm, to around 12 mm (nervellus intercepted far above centre (Fig. 7); upper tooth of mandible often weakly bidentate (Fig. 8) and scutellum frequently with spine) .... **Banchinae: Banchini** (p. 113)

–   Areolet small (e.g. Fig. 4) – if rarely, proportionately large and of rhomboidal form, then clypeus fused with face, and wing length usually much less than 7 mm (nervellus often intercepted at or below middle; mandible tridentate only in species with wing length not greater than about 5 mm; scutellum lacking spine) 4

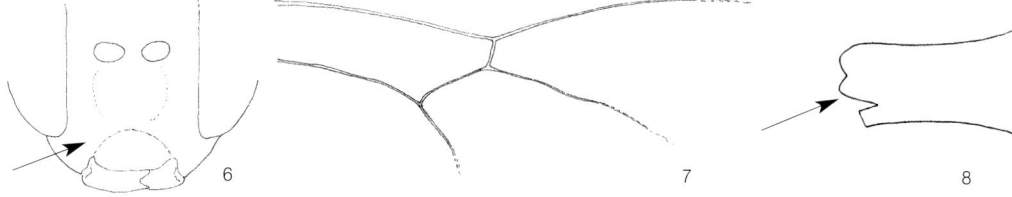

4. Wings with much of the venation non-pigmented (i.e. over distal portion of the fore wing, and most of the hind wing) .... **Neorhacodinae**

– Wings with the majority of veins strongly pigmented 5

5. Apical flagellar segment around double length of the pre-apical one; clypeus only about as wide as long (face often clothed with long silvery hair) .... **Stilbopinae**

– Apical flagellar segment usually distinctly shorter than 2 x length of pre-apical; clypeus generally much wider than long (face rarely with unusually long hair) 6

6. Mesonotum lacking coriaceous and punctate *micro*sculpture, and with deeply impressed notauli (Fig. 9a); areolet (cell 2Rs) absent; first segment of flagellum between 7-10 times longer than broad; tergite 1 of metasoma entirely rugose; male with pre-apical flagellar segments deeply excised (ovipositor lacking basal, dorsal notch) .... **Cylloceriinae**

a b
9

– Mesonotum usually with coriaceous or punctate sculpturing, the notauli weakly defined (e.g. as in Fig. 9b); areolet rarely absent; males with basal flagellar segments not excised (ovipositor often with pre-apical, dorsal notch (Fig. 10)) 7

10

7. Propodeum dorsally with only the posterior carina – this generally forming a regular arch-shaped structure (Fig. 11) (carina lateralis usually present) .... **Banchinae: Atrophini** (p. 54)

– Propodeum either with fewer, or with more dorsal carinae – or with the posterior carina quite differently formed 8

11

8. Propodeum with lateral and / or posterior carina absent or rudimentary, submetapleural carina with ventral lobe (Fig. 3a); (front tibia lacking pre-apical tooth; tarsal claws with no, or much reduced, pectin) .... **Banchinae** (a few **Atrophini**) (p. 54)

– If propodeum with much reduced carination, then submetapleural carina lacking ventral lobe; front tibia often with pre-apical tooth (Fig. 12) (tarsal claws sometimes completely pectinate) .... [*other subfamilies*]

12

Generic categories in Banchinae

In the present work, generic categories are essentially 'conservative' (including for example, *Cryptopimpla* – which arguably might rank as a mere species group within *Lissonota*). Similarly, I have continued to use '*Conoblasta*' as a subgenus (following Aubert, 1978), despite the fact that this probably cannot be supported on the basis of a wider geographical study. My reasons for maintaining the *status quo* here lie partly with preservation of a useful link with earlier literature, also with the view that changes in the status of higher groups in general should always be based on a study of the world fauna, and not that of a restricted geographic region.

Definition of species group categories

There are four kinds of classification in use at the present time: *cladistic*, based upon phylogeny; *gradistic*, also based on valid evolutionary factors, although grouping relatively primitive phyletic branches together with others manifesting strong overall similarity; *phenetic*, resembling the gradistic model in being based upon overall similarity, but not excluding convergent characteristics; and *artificial,* based upon characters chosen solely for facilitation of identification (in actual practice, 'artificial' classification will tend to carry a mosaic of cladistic, gradistic, and phenetic data). Artificial classification is especially useful (and generally entirely necessary) when dealing with large, complex taxonomic assemblages with which any attempt to impose a more natural arrangement results in the erection of unhelpful barriers to identification. This is very much the situation with the larger Banchinae genera , and particularly with species group categories within highly species-rich taxa such as *Glypta* and *Lissonota*. Townes and Townes (1978) adopted an artificial arrangement for the Nearctic *Lissonota* species, remarking that it is unlikely that proposed supra-specific taxonomic units would be applicable over a wider faunistic region. The same view applies to the majority of species group categories used in the present publication (including the separation of *Lissonota* s. str. and *Campocineta,* following Aubert, 1978). The classificatory procedure used here is thus predominantly utilitarian in nature, since the primary objective is to render identification of banchine ichneumonids as simple as possible. Since no Linnaean nomenclature is proposed for artificial categories, this approach should pose no problems for anyone wishing to probe more deeply into phylogenetic relationships within the Banchinae in the future. Finally, I have indicated where concordance with cladistic factors seems likely (and also where the artificial element lies at its most obvious).

Taxonomic objectives

In the present work, the fact that ten new species are described in a subfamily of 138 British species (around 7%) gives a good indication of the poor state of our present knowledge of the native Banchinae. Despite this comparative obscurity, a large proportion of banchines are conspicuous insects, by no means likely to be overlooked by collectors. The problem here lies mainly with the enormous number of species in the family Ichneumonidae as a whole, which, combined with unusual taxonomic difficulty, has made the family very much the province of a small number of professional entomologists with access to large collections and comprehensive library facilities. In order to understand this, it is only necessary to consider the single banchine genus *Lissonota*: over 60 species are now known in the UK, nearly double the number dealt with by Morley (1908), and the only modern literature is written in the French and Russian languages.

The present work has comprised the following main taxonomic objectives:

No previous work has attempted an adequate analysis of variation in the larger banchine genera, thus earlier keys have tended to be centred about the 'species norm', in addition often relying too heavily on colour characters. For this reason, a very high degree of success cannot be expected in applying some of Aubert's (1978) diagnoses, despite the latter being effectively the most recent revision. This deficiency has been remedied so far as possible for the British species treated in the present work.

Another serious problem with the previous literature, including both Aubert (1978) and Kuslitsky (1981), amongst recent authors, lies with the fact that data on host relationships have leant heavily on earlier publications for which both parasitoid and host identification has very often been highly suspect. In the present study, only authenticated host-rearing records are cited (unless otherwise stated).

Keys to the genera *Glypta*, *Lissonota*, *Cryptopimpla, and Exetastes* (making up the vast majority of the subfamily) are based on a full taxonomic revision, including reference to the accessible type specimens.

The creation of species groups has been necessary in the two very large genera *Glypta* and *Lissonota* – in particular, within the large subgenus *Campocineta* in *Lissonota*.

Evolutionary aspects – phylogeny

An adequate analysis of the cladistic dimension must be based upon study of the world fauna – as well as taking into account both morphological data on the larval stage and (where it exists) biochemical information. In the modern age, attention has centred on DNA sequencing, chromosome analysis, and cladistic methodology. These areas have scarcely been explored for Banchinae, although each of them will undoubtedly add a great deal to our present perspective in the future. As with any other similarly complex taxon, there will be 'biospecies' concealed beneath the anatomical curtain (just for example, with the *Glypta ceratites* complex, see p. 38).

Identification procedure

In attempting to identify banchine wasps, it is highly advisable to build at least a small collection before attempting to use the keys to species in the larger genera. Despite the fact that a reasonable proportion can be determined without going to this trouble – many species are defined by comparative traits requiring closer study. In the more taxonomically difficult sections, I would further recommend obtaining access to reliably named material (such as the collections at the Natural History Museum, London; National Museum of Scotland, Edinburgh; or Horniman Museum, London).

Identifying male specimens

It is very difficult to provide satisfactory keys to the male sex for many of the larger ichneumonid genera. There are various reasons for this. Since the problem in question is rarely discussed by authors, this can cause much frustration for non-specialists. Consequently, I provide an analysis of the situation below:

In the first place, males are frequently more variable than females. Colour characteristics tend to be subject to much variation in ichneumonids in general, and the situation is often worse with regard to the male sex. Examination of only a few males of a group of species very often uncovers 'easy' colour characters, and such differences have certainly found their way into the past literature. However, once a greater number of specimens have been examined, very many pigmentation traits prove to be of little or no value. Thus, with rare species in particular, it naturally follows that any attempt to use male colour pattern as a basis for taxonomy is prone to failure. Structural traits are also more variable in males – where even species group traits frequently manifest unexpected plasticity, and where (unlike most other insect groups), the male genitalia rarely provide much assistance in ichneumonid taxonomy.

Due to the frequently encountered phenomenon of sexual dimorphism, many males cannot be placed satisfactorily in the taxonomy without reference to verified female material (e.g. where the sexes have been reared together from a common host source). In the majority of insects, this problem is most easily solved by collecting the sexes *in cop.* but this situation is almost never encountered in the wild with ichneumonids. Male determinations based upon rearing obviously constitute the best solution to this problem. However, since many species are seldom if ever reared, it is often necessary to fall back on 'males collected with females' (i.e. at the same time and place), which it has to be admitted is a dubious second best strategy.

A further complication lies in the fact that a number of Banchines are as yet only known from one sex (in reality, some species for which males are as yet unknown may ultimately prove to be thelytokous).

Despite all of the above problems, often there will be published descriptions of reputed males, very many of which have been associated with females on the basis of pure guesswork. Consider a genus or species group of around ten closely related species. If two are unknown in the male sex, any key to males of known species will probably lead to confusion with previously unknown males. Extend this situation to genera such as *Glypta* (with 35 British species) and *Lissonota* (with over 60), and it is easy to see why difficulty is encountered with any attempt to key out the male species.

Given the above complications, it naturally follows that the majority of published keys to the male sex of large banchine genera contain areas which belong in the realm of fantasy and which are often extremely misleading, especially when leaning heavily on colour characters. My solution has been to key males wherever it seems safe to do so, then add data on others along with the female taxonomy. This approach can be very useful when dealing with reared (or simultaneously collected) material of the larger genera – *and should be used solely in that context.*

Nomenclature

Earlier workers relied upon interpretation of written descriptions, only rarely supplemented with examination of type specimens. For this reason, a great many misinterpretations have found their way into the taxonomic literature. While problems of this kind can usually be solved by obtaining access to type material, this approach can only proceed to a limited point with Banchinae. A significant proportion of the Gravenhorst banchine types were lost through dermestid damage during the Second World War – a situation which has naturally led to serious problems with nomenclature. The search for neotype material of 'lost' Gravenhorst species has been slow and painful. Although some progress has been made, a number of species names still remain with no accessible type reference material. For this reason, it has been necessary to follow the Perkins interpretations for some species. This is obviously a weak approach, in that neotypes are likely to be found which agree with Gravenhorst's overly wide description of a species – at least some of which will not be conspecific with the original type material examined by Perkins, prior to its destruction. In addition, it is unsafe to suppose that Perkins' alignment of NHM material with Gravenhorst type material was one hundred percent correct. This is hardly surprising, given that he carried out a broad analysis of type material practically throughout the Ichneumonidae in a remarkably short period of time.

The present author regards the definition of a species, along with the accumulation of information on host relationships (and biology in general) as being the essence of good taxonomy. While it is nevertheless useful to solve ongoing nomenclatural problems, the search for neotype material has been relegated to non-urgent status for the purposes of the present study.

Sensu auctorum names

There is now the question of *sensu auctorum* names, whereby a species has been long known by a name which has later been found to be based upon misidentification, or which falls into synonymy under a long forgotten prior name. It is not unusual to see widely used species names disappear completely from the literature, once a species has been correctly aligned with type material. For the purposes of the present study, I have held that widely used *sensu auctorum* names should be given prominence alongside the valid ones. This has been done by placing the names in question within square brackets

Biology

In the context of a primarily taxonomic study on banchine wasps, the principal factors discussed under the 'biology' heading will tend to centre mainly upon host preference, phenology and habitat type. Information on hosts is seen here as being of prime importance, since it is from this category of information that detailed knowledge of other aspects of the biology (including taxonomy) of a species begins to emerge. Where hosts remain unknown, good data on habitat type, phenology and geographical distribution will serve to provide initial clues for pursuit of answers to this question. It is only when these data are known that is it possible to look in more detail at parasitoid biology. In Banchinae, much of this knowledge remains somewhat fragmentary at the present time – even for several common and widely distributed species.

By way of stark contrast to a majority of the 'true' Pimplinae, banchines are koinobionts, not idiobionts. Oviposition is into early instar host larvae of Lepidoptera, and the host continues to feed and grow following that event. The parasitoid larva generally emerges from a fully grown host larva when the latter spins its cocoon or forms an underground chamber, and a lozenge-shaped silken cocoon is then formed by the ichneumonid. Overwintering may take place in the host larva or in the parasitoid's cocoon. In Glyptini, pupation tends to occur in the host's refugium.

A few banchines are of economic importance, in that they attack larvae of pest Lepidoptera. They are found in all three tribes of the Banchinae, including several which attack pests of coniferous forests, and one which oviposits into larvae of a tortricid moth found on cultivated peas.

Ecological niche

Host range in British Banchinae is somewhat limited (compared e.g. to Pimplinae), in being more or less restricted to Lepidoptera. Nevertheless, it is very diverse in the sense that a considerable variety of host pabulae are involved. Many banchine species are 'niche specialists', for example: seeking host larvae associated with either living or decaying timber (Tineidae, Oecophoridae, Sesiidae, Cossidae); attacking portable larval cases of lower lepidopteran families such as Psychidae and Adelidae; or targeting hosts that tunnel into the roots and stems of herbaceous plants (generally tortricoids or pyraloids). A number of banchines inhabit grassy habitats, attacking lepidopterous larvae feeding at stem bases or roots of Poaceae. These include several abundant parasitoid species of which very little is known of the host-parasitoid relationship. There are two reasons for this problem – firstly, many of the host species are very common, thus attract little interest from lepidopterists. Secondly, the concealed habits of host larvae render them very difficult to collect. Hunting for larvae of ubiquitous noctuids or crambids in large areas of uniform grassland is simply not a high priority with the majority of entomologists.

The niche choice situation may be broadly summarised as follows:

Tribe	Hosts
Banchini	Larvae of exophagous Lepidoptera (Noctuoidea) which pupate in soil.
Glyptini	Leaf-rolling Lepidoptera, generally Tortricoidea; also endophagous species, including hosts attacking roots, stems, and fruits. A few attack tortricid galls on conifers.
Atrophini	The *Lissonota Lampronota* group mostly attack xylophagous Lepidoptera, such as Sesiidae and Cossidae. *Lissonota* s. str. selects a wide range of host taxa, including fungivorous larvae associated with rotting timber and larvae living in soft tissues of herbaceous plants, including roots. As with Glyptini, many attack leaf-rollers (usually Tortricoidea, Yponomeutoidea, or Gelechioidea). A small proportion attack exophagous Geometrids, pyraloid larvae dwelling in mosses, case-bearing Psychidae, or tortricids forming gallular growths.

Host relationships and higher level taxonomy

Glyptini and Atrophini are predominantly parasitoids of cryptophagous or endophagous Lepidoptera. Exophagous kinds are attacked by *Cryptopimpla* and by Banchini (*Exetastes* and *Banchus*). The host-links are reflected in the form of the ovipositor of the parasitoids: Glyptini and Atrophini generally have long ovipositors, whereas *Cryptopimpla* and Banchini have the ovipositor either short or more or less concealed. In fact the schism between Banchini and the other tribes of the subfamily (being partly based upon differentials centred around form of the ovipositor) had for some considerable time concealed a proper understanding of the relationships of the *Exetastes-Banchus* assemblage to other 'classical' ichneumonid subfamilies.

Host choice and species level taxonomy

The edict that 'no two species may inhabit the same ecological niche' must not be taken to mean that no two parasitoid wasp species may attack the same host. Two different stages of the host larva may be involved, the geographical distributions of parasitoid species may differ, or even different populations of the same host feeding on different host plants may be involved. At a slightly more complex level, a common situation may be that two parasitoid species do in fact attack the same host stage, but each also attacks one or more alternative host species that are not shared by the other. In this situation, we should not be too alarmed at examples of host rearing data which appear to go against our original edict concerning parasitoid biology and ecological niche.

Distribution, abundance and phenology

A number of ichneumonid species feed as adults upon the flowers of umbellifers, which situation renders them conspicuous and easy to collect. Many species are easily taken by sweep-netting. However, the collecting and rearing of lepidopterous larvae or the employment of a Malaise trap will uncover very many species which do not frequent flower tables, and which are rarely collected by more traditional methods. In this situation, exact interpretation of the term 'rare' becomes somewhat obscured. We must also define the exact meaning of the word 'common'. The latter term has been used here, with respect to the epoch during which most collecting has taken place. While this does to a limited extent take into account the appearance of additional species in Britain through climatic warming, no attempt is made to assess the diminution of populations as a result of same (or due the effects of recently introduced agrichemicals), for the simple reason that the existing data are not sufficiently comprehensive to allow any objective judgement to be made. Nevertheless, a number of species do appear to have been of more frequent occurrence in earlier times than during the last century or so, and this situation is obviously worthy of discussion (see below).

'Historical' collections

I have used the term 'historical' in relation to a certain class of museum specimen data. 'Historical' can be approximately defined as material collected before the year 1900, for which provenance (and often also year of capture) is generally missing. 'Genuine' historical specimens have usually been associated with collectors who are known to have been active during the earlier stages of research into Ichneumonidae. The question may be asked here: can data based upon 'historical' specimens constitute evidence of a formerly greater abundance of species – or do these records merely reflect early collectors' habits? It could for example, be that these data are biased toward larger, more showy species. From the available information in Fitton's (1976) catalogue of type specimens of species described by UK authors, it would seem that there had indeed been some element of bias towards the more conspicuous species in earlier times. On the other hand, Morley (1908) cites observations by earlier authors concerning the relative abundance of large and highly conspicuous ichneumonines attacking sphingid larvae, which seem very likely to point to a genuine decrease in distribution, abundance and phenology for the species in question.

We might also ask the converse question: do certain species found more frequently during recent times reflect an increase in numbers – or is it merely that an earlier bias towards more striking forms has disappeared?

So far as a clear answer to the above questions is concerned, truly objective comparative data are lacking, and cannot be fabricated at this stage. My own impression is that some significant component of 'former abundance of rarities' is very probably due to collectors' bias towards the more conspicuous species in the past. Also that the apparent recent increase in numbers of other species is due in part to real population changes – but must also reflect the employment of collecting techniques that were unused until comparatively recent times.

Collecting

There are various means by which ichneumonids can be collected. As already indicated, the most useful modern collecting method is by use of a *Malaise trap*; very many species that had previously been considered of rare occurrence have been found to be quite common and widespread using this method. *Host-rearing* is the most valuable method of procuring not only the rarer species, but also in providing essential biological data. As with the Malaise trap method, many supposedly 'rare' ichneumonid species are quite frequently reared from host larvae – for example, the subgenus *Lampronota* of *Lissonota* (which attack Sesiidae and Cossidae), or the *Lissonota saturator* group (linked to larvae of case-bearing 'micro' Lepidoptera). Methods for accomplishing this are very much the same as those commonly used for rearing of the host species. So far as labelling of reared speciemns is concerned, this needs to be done in a methodical manner (see Fitton, Shaw, and Gauld, 1988).

In conclusion, only a limited proportion of Banchinae are likely to be found by general collecting methods such as sweeping, beating and searching blossom. Malaise trapping will invariably uncover many overlooked species in any habitat, and rearing of hosts is virtually the only method of finding many species that are very rarely encountered as free-flying adults.

Mounting

Specimens must be mounted so that all aspects of structure are fully visible. Many important characters can only be viewed in lateral or ventral aspect – thus, if direct pinning is used, then it is important that the mesosternal sulcus is not damaged, nor the mesopleural speculum obscured by the wings and / or legs. In the same way, card-pointed material must be arranged in such a way that key structures are not obscured.

Photography

Internet photographs are beginning to add a new dimension to the study of Entomology. The problem here is that a very large proportion of website images are misdetermined, and photographers are often unwilling to send voucher specimens to taxonomists. Consider the genus *Lissonota* – with over 60 British species. The largest and most distinctive of these is *L. setosa* – yet at the present time, of around half a dozen or so net images, only one is even a member of the subfamily Banchinae – and its specific identity cannot be confirmed from a photograph. It is thus absolutely essential that specimens are collected. With the exception of butterflies, most of the larger moths, and dragonflies (plus a few small sections of other orders), photographs may well supplement, but cannot possibly take the place of preserved specimens. As a general rule, working taxonomists will be always be very reluctant to offer determinations of photographs for all but a very few distinctive species. The answer to this problem is simply that photographers must either provide voucher specimens, or else expect little or no feedback from specialist taxonomists.

Checklist

This checklist summarises the conclusions of the present work. New species (sp. nov.) are in **bold** and should be attributed to Brock. New synonyms (syn nov.) are also in **bold**.

Tribe GLYPTINI

    TELUTAEA Förster, 1869
        *brischkei (*Holmgren, 1860)

    APOPHUA Morley, 1913
        bipunctoria (Thunberg, 1824)
            cubitoria (Thunberg, 1824)
            flavolineata (Gravenhorst, 1829)
            baltica (Habermehl, 1926)
        cicatricosa (Ratzeburg, 1848)
            crenulata (Thomson, 1889)
        evanescens (Ratzeburg, 1848)
            albifrons (Holmgren, 1856)
        genalis (Möller, 1883)
            superba (Hellén, 1915)

    DIBLASTOMORPHA Förster, 1869
        *cylindrato*r (Fabricius, 1787)
            erythrogaster Lucas, 1849
            bicornis Boie, 1850
            bicornis Desvignes, 1856
            corniculata Brischke, 1865
            elegans Vollenhoven, 1873
            ephippigera Kriechbaumer, 1898
            ruficornis Szépligeti, 1900
            palaeanae Kriechbaumer, 1900
            szépligetii Dalla Torre, 1901
            cylindratrix (Schulz, 1906)
            abundans (Schmiedeknecht, 1934)
        rostrata Holmgren, 1860

    GLYPTA Gravenhorst, 1829
    [*Conoblasta*]
        ceratites Gravenhorst, 1829
        elongata Holmgren, 1860
        extincta Ratzeburg, 1852
            nigriventris Thomson, 1889
         fronticornis Gravenhorst, 1829
        lapponica Holmgren, 1860
            annulata Bridgman, 1890
            areolaris Hellén, 1915
            nigricoxa (Kokujev, 1927)
            alpina (Heinrich, 1949)
        monoceros Gravenhorst, 1829

paludosa **sp. nov.**

       woerzi (Hedwig, 1952)

[*Glypta* s. str.]

       bifoveolata Gravenhorst, 1829

             setosa Roman, 1909

       consimilis Holmgren, 1860

             brevicornis Rudow, 1883

             parvicornuta Bridgman, 1886

             xanthognatha Thomson, 1889

             berolinae (Strand, 1918)

       femorator Desvignes, 1856

             filicornis Thomson, 1889

             femoratrix Schulz, 1906

             elegantula Hellén, 1915

             obscurata Kiss, 1929

             pellucida Schmiedeknecht, 1935

             triangularis Schmiedeknecht, 1935

             curvicoxa Kuslitsky, 1977

       haesitator Gravenhorst, 1829

             haesitatrix Schulz, 1906

             australis (Hedwig, 1959)

       incisa Gravenhorst, 1829

       lineata Desvignes, 1856

       longispinis (Gmelin, 1790)

             provincialis Fonscolombe, 1854

             rubicunda Bridgman, 1890

             algerica Habermehl, 1917

             zangezurica Kuslitsky, 1974

       mensurator (Fabricius, 1775)

             lugubrina Holmgren, 1860

             macropyga Hellén, 1915

             heydeni Habermehl, 1917

             jaroslavensis Shestakov, 1927

       microcera Thomson, 1889

             segrex Kokujev, 1913

       nigricornis Thomson, 1889

             rufipes Brischke, 1865

             brischkei Dalla Torre, 1901

             papyri Speiser, 1908

       nigrina Desvignes, 1856

             flavipes Desvignes, 1856

             ruficeps Desvignes, 1856

       nigrotrochanterator Strobl, 1902

             [*mensurator* auct. nec. F.]

             [*longicauda* auctt. nec. Hartig, 1838]

       parvicaudata Bridgman, 1889

       pedata Desvignes, 1856

             varicoxa **Thomson, 1889 syn. nov.**

       pictipes Taschenberg, 1863

       punctifrons Bridgman, 1889

pusilla **sp. nov.**
　　　　　(*scalaris* auctt. nec Grav.)
resinanae Hartig, 1838
　　　　arreptans Hellén, 1915
　　　　summimontis Heinrich, 1953
rufata Bridgman, 1887
sculpturata Gravenhorst, 1829
　　　　macrura Habermehl, 1918
　　　　rufoclypeata Kiss, 1924
scutellaris Thomson, 1889
similis Bridgman, 1886
　　　　rufipes Thomson, 1889
　　　　thomsonii Dalla Torre, 1901
　　　　thomsoni Strobl, 1902
tenuicornis Thomson, 1889
　　　　pygmaea Shestakov, 1927
trochanterata Bridgman, 1886
ulbrichti Habermehl, 1926
vulnerator Gravenhorst, 1829
　　　　vulneratrix Schulz, 1906
　　　　monstrosa Hellén, 1915

Glyptini erroneously recorded as British / Irish:
　　　　scalaris Gravenhorst, 1829
　　　　schneideri Krieger, 1897
　　　　teres Gravenhorst, 1829
　　　　gracilis Hellén, 1915

Tribe ATROPHINI
　SYZEUCTUS Förster, 1869
　　　　bicornis (Gravenhorst, 1829)
　　　　fuscator (Panzer, 1809)
　　　　　maculatorius (Fabricius, 1787)
　　　　　bicolor Szépligeti, 1889
　　　　　rufipes Kiss, 1933

　ARENETRA Holmgren, 1859
　　　　pilosella (Gravenhorst 1829)

　ALLOPLASTA Förster, 1869
　　　　piceator (Thunberg, 1822)
　　　　　creditor (Thunberg, 1822)
　　　　　albitarsus (Gravenhorst, 1829)
　　　　　lata (Gravenhorst, 1829)
　　　　　murina (Gravenhorst, 1829)
　　　　　albitarsoria (Zetterstedt, 1838)
　　　　　genucincta (Rudow, 1886)
　　　　　variipes (Szépligeti, 1899)
　　　　plantaria (Gravenhorst, 1829)

LISSONOTA Gravenhorst, 1829
[*Meniscus*]
    lineolaris (Gmelin, 1790)
        catenator Panzer, 1804
        gladiator Thunberg, 1824
        mammilator Thunberg, 1824
        signator Thunberg, 1824
        excavator Thunberg, 1824
        facialis Desvignes, 1862
        sachalinensis Matsumara, 1911
[*Lampronota*]
    biguttata Holmgren, 1860
        femorata Holmgren, 1860
        crassipes Thomson, 1877
    canaliculata (Szépligeti, 1899)
        ***pimplator* auctt. nec Zetterstedt, 1838 [syn nov.]**
        flavipes Lucas, 1849
    deversor Gravenhorst, 1829
    ***dormitor* sp. nov.**
    freyi (Hellén, 1915)
        tuberculata (Hellén, 1915)
        sesiae Habermehl, 1918
    frontalis (Desvignes, 1856)
        sulcator (Morley, 1908)
    fulvipes (Desvignes, 1856)
        piffardi (Morley, 1908)
    nitida Gravenhorst, 1829
        agnata Gravenhorst, 1829
        rhenana Ulbricht, 1916
        lissonotoides (Habermehl, 1917)
    pimplator (Zetterstedt, 1838)
    ***plana* sp. nov.**
        (*impressor* auctt. nec Grav)
    setosa (Geoffroy, 1785)
        enervator (Fabricius, 1793)
        cryptator (Thunberg, 1824)
        removator (Thunberg, 1824)
        nigra (Szépligeti, 1914)
[*Lissonota*]
    clypeator (Gravenhorst, 1820)
        cylindrator auctt. nec F., 1787
        unicornis Strobl, 1902
        nigrescens Constantineanou, 1929
        spectacabilis Schmiedeknecht, 1935
        magma Hienrich, 1952
    digestor (Thunberg, 1824)
        vocator (Thunberg, 1824)
        hians Thomson, 1877

fundator (Thunberg, 1824)
 sulphurifera Gravenhorst, 1829
 rimator Thomson, 1877
 affinis Szépligeti, 1899
 caudata (Szépligeti ,1899)
 ruficoxis (Schmiedeknecht, 1900)
 nigricoxis Pfankuch, 1920
impressor Gravenhorst, 1829
 basalis Brischke, 1865
 signata (Szépligeti, 1899)
 nigricoxis Ulbricht, 1913
 humerella Habermehl, 1918
magdalenae Pfankuch, 1921
 vernalis Roman, 1925
sabulosa sp nov.
[*Loxonota*]
cruentator (Panzer, 1809)
 insignita Gravenhorst, 1829
 verberans Gravenhorst, 1829
 cruentatrix (Schulz, 1906)
 rufifemur Kiss, 1926
histrio (Fabricius, 1798)
 parallela Gravenhortst, 1829
 dioszhegyi (Kiss, 1924)
 nigrobasalis Constantineanu and Pisica, 1960
lineata Gravenhorst, 1829
[*Campocineta*]
accusator (Fabricius, 1793)
 rusticator (Thunberg, 1824)
 humeralis (Zetterstedt, 1838)
 unicincta Holmgren, 1860
 thomsoni Schmiedeknecht, 1900
 nigricoxa Strobl, 1902
 segmentator auctt. nec Grav.
 segmentellator Aubert, 1967
admontensis Strobl, 1902
 alpina Strobl, 1902
 praebellator Aubert, 1967
antennalis Thomson, 1877
arborator sp. nov.
argiola Gravenhorst, 1829
 eximia Habermehl, 1918
?buccator (Thunberg, 1824)
carbonaria Holmgren, 1860
 melania Holmgren, 1860
 artemesiae Tschek, 1871
clypealis Thomson, 1877
 albobarbata Strobl, 1902
consobrina sp. nov.

coracina (Gmelin, 1790)
      bellator (Gravenhorst, 1807)
      tricoloria (Thunberg, 1824)
      irrigua Thomson, 1888
      bellatrix Schulz, 1906
      meridionalis Seyrig, 1928
culiciformis Gravenhorst, 1829
      lateralis Gravenhorst, 1829
      cruenta Vollenhoven, 1858
      assimilis Brischke, 1880
      sziladii Kiss, 1926
dubia Holmgren, 1856
      jugorum (Strobl, 1903)
      duplanae (Heinrich, 1937)
erythrina Holmgren, 1860
      pusilla Habermehl, 1918
fletcheri Bridgman, 1882
folii Thomson, 1877
      transversa Bridgman, 1889
      areolata Kiss, 1924
genator Aubert, 1972
gracilenta Holmgren, 1860
gracilipes Thomson, 1877
halidayi Holmgren, 1860
linearis Gravenhorst, 1829
      varicornis (Schmiedeknecht, 1900)
      incerta Habermehl, 1918
luffiator Aubert, 1969
maculata Brischke, 1865
      affinis Brischke, 1865
mutator Aubert, 1969
nigridens Thomson, 1889
obsoleta Bridgman, 1889
palpalis Thomson, 1889
      oudemansi Smits van Burgst, 1912
      exareolata (Habermehl, 1923)
      inareolata (Kiss, 1924)
palpator Aubert, 1969
      ***parasitellae* Horstman, 2003 syn. nov.**
picticoxis Schmiedeknecht, 1900
pleuralis Brischke, 1880
      strigifrons Schmiedeknecht, 1900
proxima Fonscolombe, 1854
      varipes (Desvignes, 1856)
      commixta Holmgren, 1860
      lapponica Holmgren, 1860
      variipes Dalla Torre, 1901
punctiventrator Aubert, 1977
punctiventris Thomson, 1877
      ?errabunda Holmgren, 1860

quadrinotata Gravenhorst, 1829
  leucogona Gravenhorst, 1829
  carinifrons Thomson, 1877
saturator (Thunberg, 1824)
  basalis Thomson, 1889
semirufa (Desvignes, 1856)
***serena* sp. nov.**
***simulator* sp. nov.**
stigmator Aubert, 1972
subaciculata Bridgman, 1886
  nitida Bridgman, 1886
tenerrima Thomson, 1877
  ***variabilis* Holmgren, 1860 syn. nov.**
trochanterator Aubert, 1972
versicolor Holmgren, 1860
  formosa Bridgman, 1888
  coxator Smits van Burgst, 1914
  rufithorax Habermehl, 1918
***virgata* sp. nov.**

CRYPTOPIMPLA Taschenberg, 1863
  altipes (Holmgren, 1860) stat. nov.
  anomala (Holmgren, 1860)
  arvicola (Gravenhorst, 1829)
    brachycentra (Gravenhorst, 1829)
    kaisdii (Kiss, 1929)
  calceolata (Gravenhorst, 1829)
    leptogaster (Holmgren, 1860)
  caligata (Gravenhorst, 1829)
  errabunda (Gravenhorst, 1829)
  hertrichi Heinrich, 1952
  quadrilineata (Gravenhorst, 1829)
    blanda (Gravenhorst, 1829)
  sp. indet.

Atrophini erroneously recorded as British / Irish:
  Syzeuctus irrisorius Rossi, 1794
  Lissonota silvatica Habermehl, 1918

Tribe BANCHINI
  EXETASTES Gravenhorst, 1829
    adpressorius (Thunberg, 1824)
      guttatorius Gravenhorst, 1829
      tristis Gravenhorst, 1829
      procera Kriechbaumer, 1894
      guttifer Thomson, 1897
      medianus Szépligeti, 1898
      albopictus Aubert, 1959
      albopictor Aubert, 1972

atrator (Förster, 1771)
    cinctipes (Retzius, 1783)
    junci (Geoffroy, 1785)
    osculariatus (Fabricius, 1787)
    obscurator (Gmelin, 1790)
    clavator (Fabricius, 1793)
    tarsator (Fabricius, 1804)
    sinuatorius (Thunberg, 1824)
calobatus Gravenhorst, 1829
    calobates Dalla Torre, 1901
femorator Desvignes, 1856
fornicator (Fabricius, 1781)
    expansor (Thunberg, 1824)
    punctulatus Kokujev, 1905
illusor Gravenhorst, 1829
    minor Szépligeti, 1901
    annulatus Habermehl, 1927
    ?geniculosus Holmgren, 1860
illyricus Strobl, 1904
*laevigato*r (Villers, 1789)
    cothurnatus (Gravenhorst, 1807)
    incurvator (Thunberg, 1824)
    alpinus Kriechbaumer, 1888
    puberulus (Szépligeti, 1898)
    levigator Dalla Torre, 1901
    similis Kokojev, 1905
    nigriventris Meyer, 1978
maurus Desvignes, 1856
    facialis Desvignes, 1856
    benoisti Seyrig, 1926
    melanopus Meyer, 1927
    croaticus Hensch, 1928
nigripes Gravenhorst, 1829
tibialis Pfankuch, 1921

BANCHUS Fabricius, 1798
    crefeldensis Ulbricht, 1916
        croaticus Hensch, 1928
    dilatatorius (Thunberg, 1824)
        acuminator (Fabricius, 1787)
        compressus (Fabricius, 1787)
        sibiricus Meyer, 1927
    falcatorius (Fabricius, 1775)
        variegator (Fabricius, 1775)
        tricolor (Schrank, 1776)
        intersectus (Geoffroy, 1785)
        aries (Christ, 1791)
        notatorius (Olivier, 1792)
        histrio (Schrank, 1802)
        labiatus (Schrank, 1802)

falcator Fabricius, 1804
luteofasciatus Ulbricht, 1911
nobilitator Morley, 1915
sanguinator Morley, 1922
lavrovi Meyer, 1927
nigromarginatus Constantineanu and Pisica, 1960
propitus Kuslitsky, 1979
hastator (Fabricius, 1793)
pungitor (Thunberg, 1824)
reticulator (Thunberg, 1824)
femoralis Thomson, 1897
kolosovi Meyer, 1925
moppiti Fitton, 1985
palpalis Ruthe, 1859
pictus Fabricius, 1798
cultratus (Gmelin, 1790)
mutillatus (Christ, 1791)
bipunctatus Hensch, 1928
zagoriensis Hensch, 1928
volutatorius (Linnaeus, 1758)
venator (Linnaeus, 1758)
umbellatorum (Schrank, 1786)
certator (Thunberg, 1824)
monileatus Gravenhorst, 1829
farrani Curtis, 1836
moniliatus Marshall, 1872
alticola Schmiedeknecht, 1910
calcaratus Szépligeti, 1910

RYNCHOBANCHUS Kriechbaumer, 1894
flavopictus Heinrich, 1937

Key to tribes of Banchinae

1. Tergites 2-3 with a pair of diagonal grooves defining an embossed area (Fig. 13); areolet usually absent; genal carina usually meeting hypostomal carina at mandible base (Fig. 14) .... **Glyptini** (p. 26)

–   Tergites without oblique grooves and embossed regions; areolet usually present (Fig. 17); genal carina usually meeting hypostomal behind mandible base (Fig. 15) 2

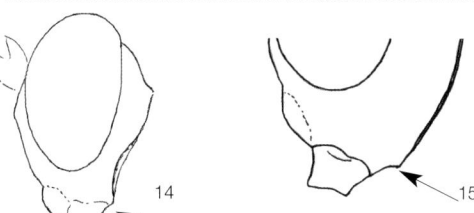
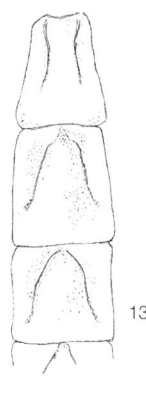

2. Nervellus in hind wing broken near (usually below) centre (or else second section of cubitus absent) (Fig. 16); areolet small (Fig. 17) or incomplete; metasoma not laterally compressed; ovipositor long (often longer than metasoma); clypeal margin evenly curved (Fig. 18); (propodeal spiracle usually subcircular – sometimes moderately elongate (Fig. 19); scutellum never with apical spine) .... **Atrophini** (p. 54)

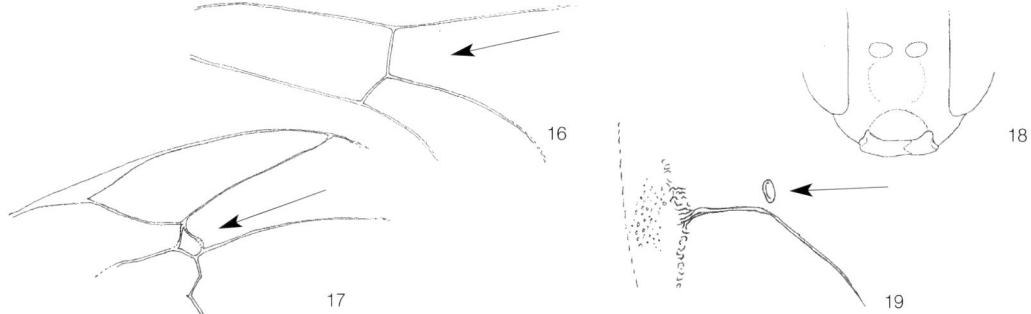

–   Nervellus in hind wing broken far above centre (Figs 20, 21) areolet large and rhomboidal (Fig. 22); apex of metasoma very often laterally compressed, at least in females; ovipositor concealed, to shortly projecting; clypeal margin not evenly curved – either truncate, thus with the central part straight-edged – or else with central emargination, rounded laterally (Fig. 23) (propodeal spiracle up to over 4 times longer than broad, scutellum often with apical spine (e.g. Fig. 24)) .... **Banchini** (p. 113)

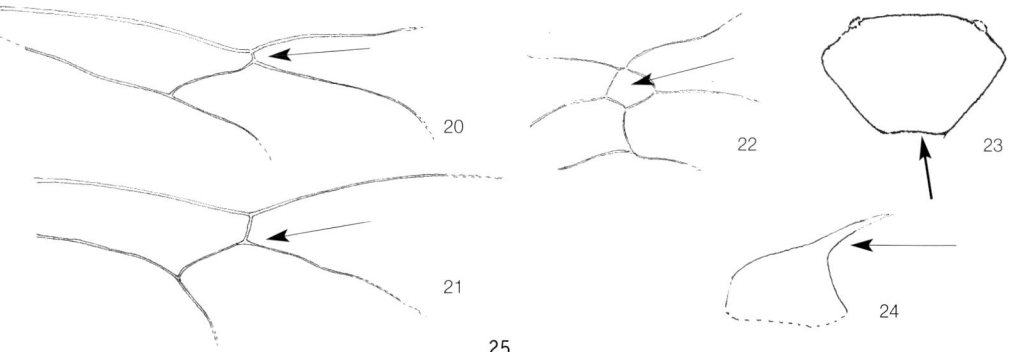

Tribe Glyptini

The Glyptini is certainly one of the most easily recognisable higher group taxa in the Ichneumonidae. Members of the tribe mostly attack leaf-rolling or internal tissue-feeding Tortricid larvae.

Only *Glypta haesitator* is listed as Holartic in the Dasch revision (1988) and this purely in the context of an introduction linked to pest control. Around a dozen native Nearctic species are reported as important parasitoids of pest Lepidoptera.

Keys to the genera/subgenera of Glyptini

1.   Frons with two tubercles that are fused basally to the antennal sockets (Figs 25, 26); areolet present; genal carina meeting hypostomal carina well above mandible base; propodeum with only the apical transverse carina (cf. Fig. 27); clypeus with a thin apical flange that is widest laterally (thus appearing emarginate at centre) (Fig. 28) .... ***Telutaea*** (p. 28)

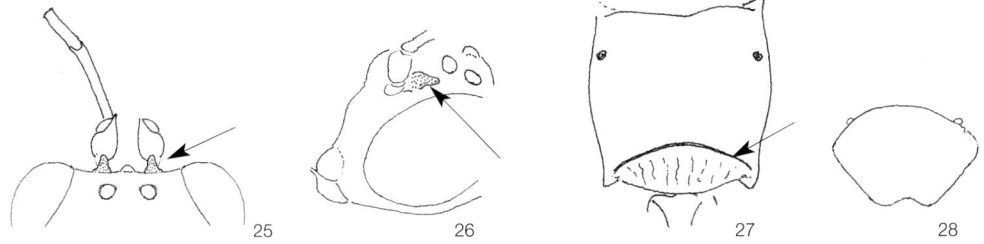

−   If frons has a pair of tubercles, these are free from the antennal sockets (Fig. 29); areolet absent (Fig. 30); genal carina joining hypostomal at base of mandible (cf. Fig. 31); propodeum usually with lateral and or dorsal carinae (Fig. 32); clypeus with no apical flange, rounded apically 2

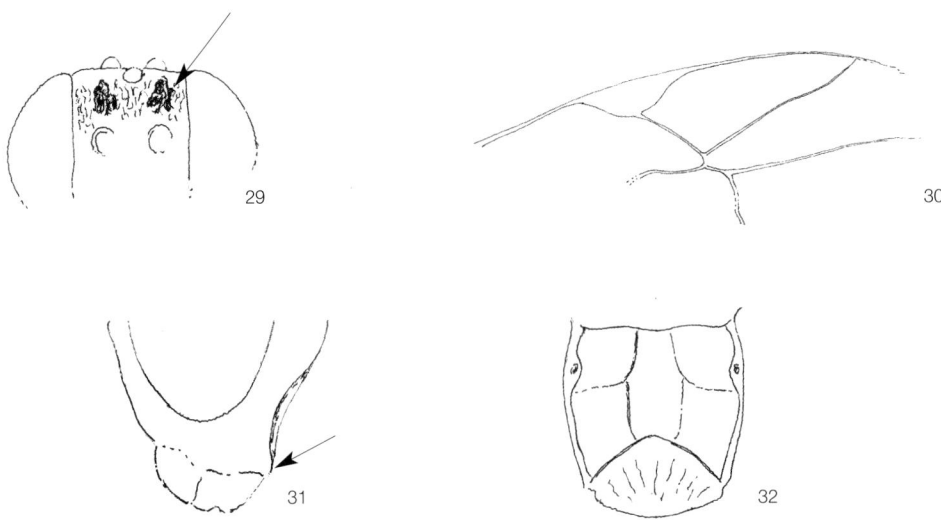

2. Genal carina strongly sinuate, not interrupted, the postgena twisted onto genal plane (Fig. 33); spur of front tibia longer than 0.5 x length of barsitarsus, distinctly sinuate (Fig. 34); tergites 2-4 often with a central carina; tergite 2 striate in front between the diagonal impressions (Fig. 35) – excepting when tergite 1 about 2 times longer than broad .... **Apophua** (p. 28)

– If the genal carina is strongly sinuate, it is largely erased (Fig. 36); postgena never twisted onto genal plane; spur of front tibia shorter than 0.5 x barsitarsus length (often only weakly sinuate); tergites 2-4 lacking central carina – tergite 2 never striate between the diagonal impressions 3

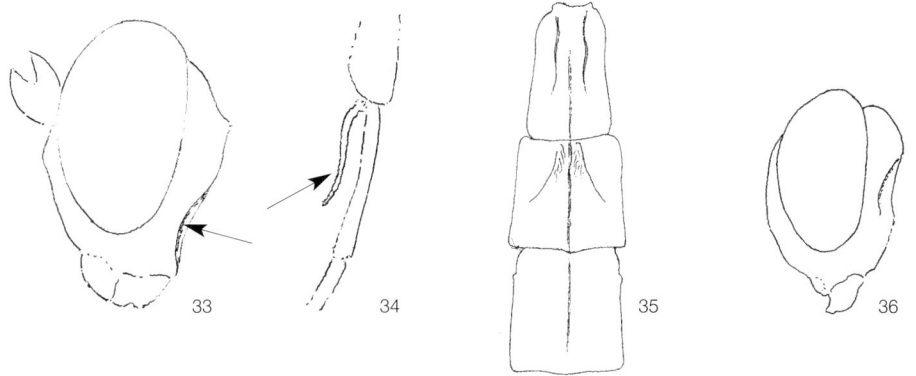

33   34   35   36

3. Malar space about 2 x mandible width and a little less than, to around maximum, temple length (Fig. 37); frons heavily crenulate at centre (cf. Fig. 29); flagellum 1 over 6 x longer than broad; frontal horn double when present (Fig. 38); speculum very small (as Fig. 39) ....
.... **Diblastomorpha** (p. 31)

37   38   39

– Malar space rarely approaching 2 x mandible width (Fig. 33) and much shorter than maximum temple length; frons coriaceous, punctate, or striate centrally; flagellum 1 usually much less than 6 x longer than broad; frontal horn single when present (Fig. 40); speculum often large (Fig. 41), sometimes absent .... [**Glypta** sensu lato] ... 4

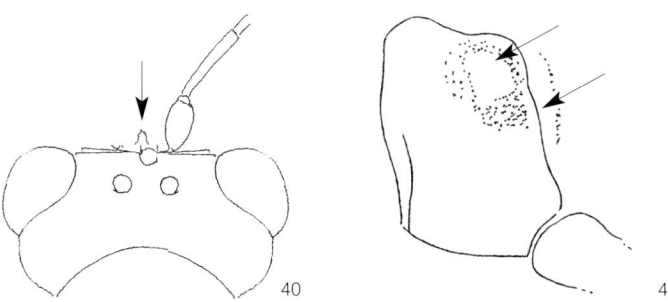

40   41

4. Inter-antennal horn present, hollow dorsally and usually blade-like ventrally – with or without lateral convolutions; frons generally impunctate medially (Fig. 42) .... subgenus ***Conoblasta*** (p. 32)

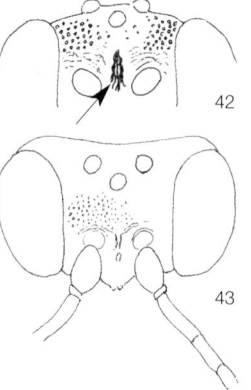

42

–   Inter-antennal horn absent (if a low inter-antennal crest is present, then with frons punctate medially as well as laterally (Fig. 43) .... subgenus ***Glypta*** (p. 38)

43

Genus *Telutaea* Förster, 1869

This genus is easily recognised – however, the frontal horns of the sole British species, *T. brischkei* do not constitute a generic character.

–   Antennal scrobes strongly raised dorsally; frons entirely glabrous; interocellar space only about 0.5 x ocellus to occipital carina; tergites 1-4 strongly elongate (fore wing length about 6 mm). *Rare* .... ***brischkei*** (Holmgren)

Species account for *Telutaea*

Telutaea brischkei (Holmgren, 1860) Plate 1

Taxonomy: *Telutaea* is unlikely to be mistaken for any other banchine genus (note that the pair of frontal horns in the common and widely distributed *Diblastomorpha* is not connected to the antennal scrobes).

Distribution, abundance and phenology: Little known in the UK. Specimens have been taken during June and July, but also October – possibly indicating presence of two generations.

England: 6 female, 8 male, Oxfordshire, Taynton Fen, Malaise trap; Berkshire, Woolhampton, ex cocoon on *Urtica*, 21.vi.1979, emerged 30.vi.79; Norfolk, Catfield, 3.x.1983 [NMS].

Biology: The fragmentary distribution data at hand may imply a link to fenland. Aubert (1978) cites a rearing from an *Archips* (*Cacoecia*) species which does not occur in the UK. The indigenous rearing given above may imply a link to herbaceous plants.

Genus *Apophua* Morley, 1913

The commoner *Apophua* species have been reared from Tortricidae on foliage of deciduous trees. Accordingly, they are mostly to be found in deciduous woodland. Aubert (1978) lists only the four species treated here for the Western Palaearctic, while Dasch (1988) reports a single species occurring in the Nearctic region.

Fore wing lengths lie in the range of 5-6 mm.

Key to species for *Apophua*

1. Minimum genal length around 0.66 x width of mandible base (Fig. 44); malar space nearly 2 x width of mandible base; clypeus and mandibles dark testaceous (pale-marked in males); hind femur and tibia reddish – latter with dark apex; ocellus to eye distinctly greater than interocellar distance in females; mesopleurum strongly punctured (Fig. 45a); tergal impressions subrectangular, tergite 3 strongly transverse (Fig. 46). *Rare* .... *genalis* (Möller) (p.30)

–   Minimum genal length at most 0.3 x width of mandible base (Fig. 47) malar space narrower than mandible base; clypeus and mandibles bright yellow; hind femur with dark apex, hind tibia with pale base; ocellus to eye at most equal to interocellar distance; mesopleurum finely punctured (Fig. 45b); (tergal impressions usually strongly acute (Fig. 48); tergite 3 transverse, subquadrate or elongate) 2

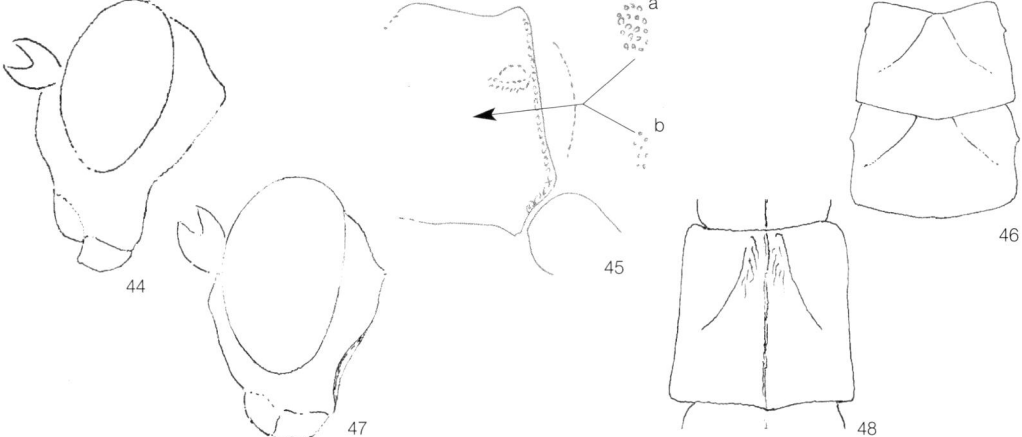

2. Propodeum lacking areae enclosed by carinae (Fig. 49), tergite 1 long and slender; dorsal areas of pronotum black, with yellow mark adjacent to tegulae; mesonotum and metasoma finely punctate; tergite 3 longer than broad. *Common* .... *evanescens* (Ratzeburg) (p. 30)

–   Propodeum with areae (Fig. 50), tergite 1 less than 2 x longer than broad; pronotum usually with yellow shoulder stripe running to tegula; mesonotum and metasoma strongly punctate; tergite 3 transverse to slightly longer than broad 3

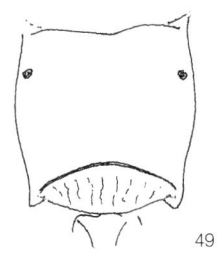

3. Tergites 2-3 with a strong central carina; basal transverse carina of propodeum absent; tergite 3 subquadrate to longer than broad, the tergal impressions tending to be more acute. *Common* .... *bipunctoria* (Thunberg) (p.30)

–   Tergites 2-3 with vestigial central carina; basal transverse carina of propodeum present (Fig. 50); tergite 3 subquadrate to broader than long, the impressions subrectangular. *Rare* .... *cicatricosa* (Ratzeburg) (p.30)

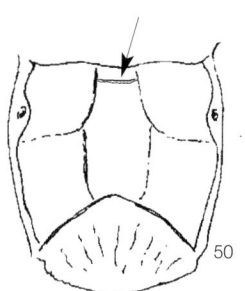

Species accounts for *Apophua*

Apophua genalis (Möller, 1883)

Taxonomy: The abnormally large malar space renders this species easy to recognise. In males the ocellar interspace is similar to that between ocellus and eye.

Distribution, abundance and phenology: Few records from the UK. England: one female, no data (Desvignes) [NHM]; more recently: one female, Cornwall, Redmoor NR, 1983; Yorkshire, Malham Tarn, 1958 [MU]. Wales: Borth Bog, Cardiganshire, 1985 [NMS]. Ireland: Armagh; Donegal; Down (O'Connor *et al.*, 2007). Flight period: has been collected during July.

Biology: There is a rearing record from *Aphelia viburnana* (D. & S.) [Bridgman Collection, NCM]. It is difficult to predict habitat preference on the basis of the scant existing data, but the species does not appear to possess the usual woodland association found in other members of the genus.

Apophua evanescens (Ratzeburg, 1848)

Taxonomy: The distinctive metasomal shape and sculpturing should leave no difficulty in identifying the species.

Distribution, abundance and phenology: Common and widespread, although apparently becoming more scarce towards the north (records extend to Llanwyrtyd, Wales, and Ayrshire, Scotland). Apparently of southern distribution in Ireland: Wicklow; Waterford; Wexford (O'Connor *et al.*, 2007). Flight period: July to October.

Biology: *Glypta evanescens* is clearly a species of deciduous woodland. Hosts are Tortricidae on *Quercus* (reared from *Pandemis corylana* (F.) [NHM] and *P. cinnamomeana* (Treits.) [NMS]). Sometimes attracted to light traps.

Apophua bipunctoria (Thunberg, 1822)

Taxonomy: Readily distinguished from *genalis* and *evanescens* but needs careful discrimination from *cicatricosa*.

Distribution, abundance and phenology: Widespread and not uncommon, although apparently becoming less frequent northwards (occurring at least as far north as Aberdeenshire). Ireland: records extend to Clare and Wicklow (O'Connor *et al.*, 2007). Flight period: June and July.

Biology: *A. bipunctoria* has been reared from Tortricidae on *Quercus,* including *Pandemis cerasana* (Huebn.). England: Oxfordshire, Wytham Wood [NHM] – also from the same host on *Lonicera*. Other hosts: '?*Apotomis betuletana*' (Haw.); '?*Epiphyas postvittana* (Walk.)' on *Prunus*; *Adoxophyes orana* (Fisch. v. Roesl.) on *Quercus*; indet. host, *Salix*; '*Pandemis cinnamomeana* (Treits.) or *Spilonotus occelana*'; *Ptycholoma lecheana* (L.) on *Alnus*; '?*Acleris variegana*' (D. & S.) [NMS]. As with *evanescens*, the biotope is deciduous woodland.

Apophua cicatricosa (Ratzeburg, 1848) Plate 2

Taxonomy: The present species has previously been distinguished from *bipunctoria*, largely on the basis of unreliable characters (it would be useful to see further material in order to fully assess species boundaries).

Distribution, abundance and phenology: Few British specimens in collections. England: 'Desvignes'; Surrey, Bookham; two specimens from Morley Coll. ('Badlaws' / 'Cross B') [NHM]. One female, no data [NCM]. Flight period: has been taken at M.V. light during July.

Biology: The only reared specimen that I have personally encountered was reputedly reared from *Archanara geminipunctata* (Noctuidae) – which can only be a misidentification of host.

Genus *Diblastomorpha* Förster, 1869

Diblastomorpha could arguably be treated as a subgenus of *Glypta*. The taxon is easily recognised from characters given in the key (fore wing lengths around 6 mm).

Key to species for *Diblastomorpha*

1. Frons with two horns (Figs 51, 52). *Common and generally distributed* .... *cylindrator* (Fabricius)

–   Frons with no horns. *Rare (possibly less so in Ireland)* .... *rostrata* Holmgren

51

52

Species accounts for *Diblastomorpha*

Diblastomorpha cylindrator (Fabricius, 1787) [*bicornis*] Plate 3

Taxonomy: Due to the distinctive frontal horns, this species is unlikely to be confused with any other member of the tribe.

Distribution, abundance and phenology: Very widespread in the UK, and generally of common occurrence. Ireland: widespread (O'Connor *et al.*, 2007). On the wing during July and August.

Biology: *G. cylindrator* has frequently been reared from *Aphelia paleana* (Huebn.). Additional hosts data: *Cnephasia* sp. on *Onobrychis* [NHM]; ex *Agrimonia* leaf roll; ex *Aphelia unitana* (Huebn.) [NMS]. The species occurs on open country, including hedgerows, commons, waste ground and similar habitats.

Diblastomorpha rostrata Holmgren, 1860

Taxonomy: *G. rostrata* has often been treated as being conspecific with *cylindrator*. However, it seems not to have been collected concurrently with *cylindrator*. Due to the small contingent of specimens available for study, its status remains somewhat uncertain.

Distribution, abundance and phenology: Ireland: The Maurough, 'S. of Newcastle'; Wexford, Rosslare, 1963; Isle of Man: The Curraghs, Goshen. There are few British Islands mainland records: three females, two males, England, Dorset [NHM]; two others, labels illegible [NMS]. Dates of capture: June-August.

Biology: Reared from *A. paleana* (Huebn.) [NMS]. The habitat may be similar to that of the preceding species.

Genus *Glypta* s. str. Gravenhorst 1829

Aubert (1978) listed around 60 *Glypta* species for the Western Palaearctic region, and Kuslitsky (1981) treated 56 in the European part of the former USSR. Dasch (1988) treated 311 Nearctic species. Host records cited by the latter author include Lasiocampidae, Geometridae, Lycaenidae, Noctuidae – also Cephidae and Coleoptera (Scarabaeidae, Cerambycidae, Rhyncophora). Some proportion of these 'extraneous' hosts are probably based upon misidentification. Nevertheless, it may be of significance that there are reputedly authentic rearings of the Nearctic *G. saperdae* from *Saperda* (Cerambycidae), from two different localities.

My own taxonomic treatment follows Aubert in recognising *Conoblasta* for the species group centred on *G. ceratites* (however, see remarks under the *Conoblasta* heading).

Glypta subgenus *Conoblasta* Foerster 1868

British *Conoblasta* species appear to form a compact natural group. However, the validity of the subgenus has to be weighed against the considerable variety of form found in the frontal horn and associated structures in many Nearctic species. Dasch (1988) lists *Conoblasta* as a synonym of *Glypta*, without comment.

This is the most difficult species complex within Banchinae. The key given here is in part simplified (see comments on intraspecific variation within the *C. ceratites* aggregate). In this context, Kuslitsky (1981) uses colour characters that are not reliable for British material, also some aspects of frontal horn morphology which are certainly variable within a species. Ovipositor length in this group varies between length of metasoma to full body length – usually lying somewhere between these two extremes. Variation within a species is too great to allow use of ovipositor length as a taxonomic marker. Besides the structural traits given in the key, extensive reddish colouration (i.e. over at least one abdominal tergite) is more frequently encountered (by virtue of relative abundance in the field) in species within the *elongata* group, than it is in females of species akin to *ceratites*. Finally, it must be said that *G. paludosa* lies somewhat intermediate between the *elongata* and *ceratites*-related sections of the subgenus. Consequently, it is keyed within both domains.

Key to species for subgenus *Conoblasta*

1. Lateral contour of temple convex in dorsal view – maximum temple length at least 0.8 x first flagellar segment, latter up to a little greater than 5 x longer than broad (Fig. 53); (central tergites subquadrate to longer than broad; hind tibia often with only the apical, or no dark annules; hind tarsal segments frequently reddish-fuscous with paler bases). Species found predominantly in marshland, fen and damp woodland 2

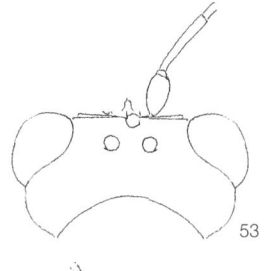
53

–   Lateral contour of temple almost straight-edged in dorsal view – maximum temple length at most 0.8 x first flagellar segment, latter up to nearly 6 x longer than broad (Fig. 54); (central tergites subquadrate to broader than long; hind tibiae usually with sub-basal and apical dark bands, whitish basally; tarsal segments blackish with white bases). Species of forest, heath, and common, not restricted to wetlands 4

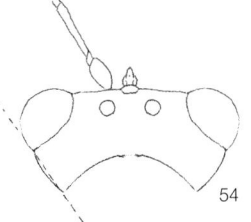
54

2. Temple only weakly convex (Fig. 56); tergite 3 (plus usually 2) distinctly longer than broad – both piceous in colour; flagellum 2 about 2.5 x longer than broad; metasoma entirely blackish, tibia 3 pale testaceous, whitish at base, and with sub-basal and apical dark annules, tarsus 3 segments dark with pale base; temporo-genal region usually biplanar (Fig. 55); ventral keel of horn usually moderately convoluted (smaller species, fore wing length 5-5.5 mm). *Uncommon, but widespread, including moist forest clearings, in addition to wetlands* .... ***paludosa* sp. nov.** (p. 35)

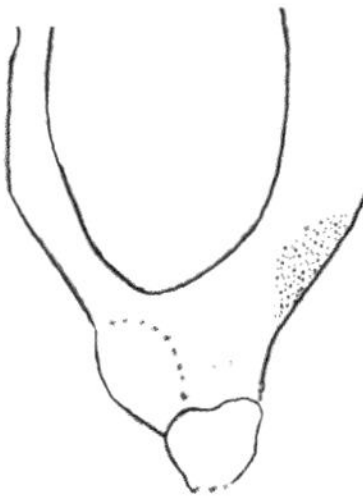

55

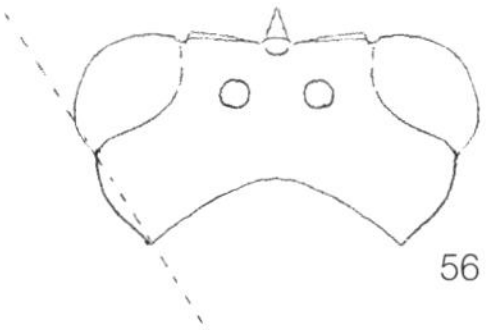

56

– Temple more distinctly convex (Fig. 53) – excepting sometimes when central tergites red; tergites 2 and 3 subquadrate to slightly longer than broad; flagellum 2 nearer 3 x longer than broad; metasoma somtimes extensively red-marked; tibia 3 reddish, usually darkened at apex alone, tarsus 3 predominantly reddish; temporo-genal region often evenly convex (ventral keel of horn often strongly convoluted (Fig. 57). Usually larger, fore wing length up to 6.5mm). *Wetland species* 3

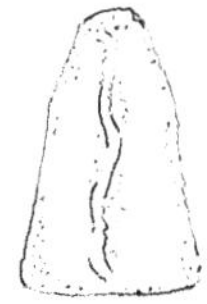

57

3. Metasoma usually predominantly dark, rarely extensively reddish-suffused over central tergites; temples strongly convex in dorsal aspect; temporo-genal region evenly convex (horn with ventral keel slightly, to strongly convoluted). *Common and widely distributed in wetlands* .... ***elongata*** Holmgren (p. 36)

– Metasoma usually with red central tergites; temples more weakly convex in dorsal aspect; temporo-genal region often weakly biplanar (see Fig. 55); horn with ventral keel straight, often blade-like (Fig. 58). *Marshland, uncommon* .... ***monoceros*** Grav. (p. 36)

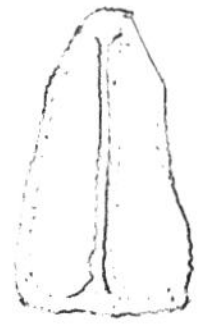

58

4. Large species (fore wing length at least 7 mm); ventral surface of horn with two well-defined lateral carinae (Fig. 59); (costula as strong, or stronger than other propodeal carinae; frons and lower temple from moderately to strongly punctured (Figs 60, 61); metasoma entirely black). *Rare (probably overlooked)* .... ***woerzi*** (Hedwig) (p. 36)

– Smaller species (fore wing length at most 6 mm); ventral surface of horn either smooth laterally, or with wrinkled convolutions (costula frequently absent or ill-defined; frons and lower temple often sparsely punctured (Fig. 62), or with little or no puncturation; metasoma sometimes conspicuously red-marked) 5

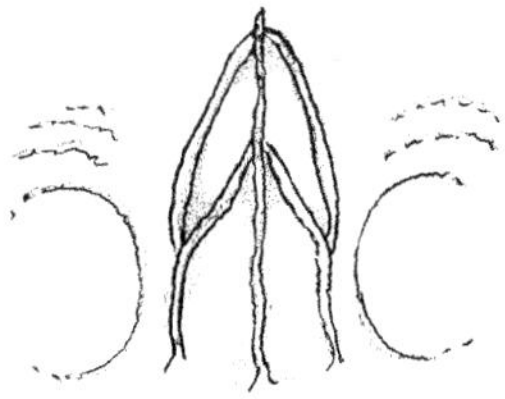

59

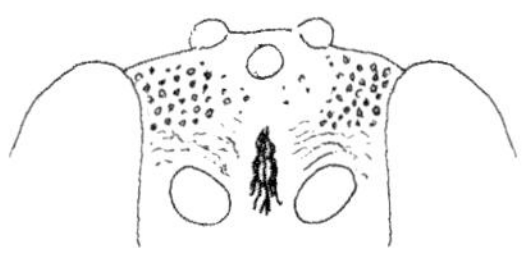

60

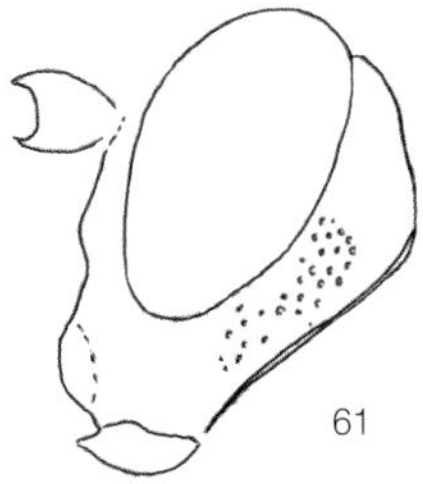

61

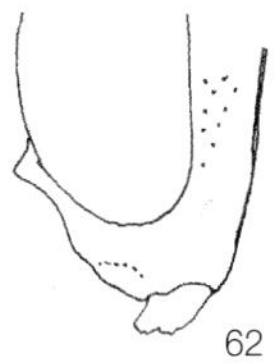

62

5. Horn with knife-edged ventral keel, lacking distinct lateral swellings, often forming a strongly acute-angled triangle in dorsal view, the dorsal cavity parallel-sided to weakly spatulate; tibia 3 testaceous at base, female metasoma with central tergites having broad apical red bands (frons laterally, with strong punctures, approximately same diameter as interspaces; temporo-genal region distinctly punctate). *Uncommon* .... **fronticornis** Grav. (p. 37)

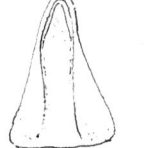

63

– Horn often swollen and convoluted laterally, less acutely angled in dorsal aspect, the dorsal cavity spatulate (Fig. 63); tibia 3 yellowish or whitish at base; if metasoma red-marked in females, then not in regular subapical bands (frons often sparsely punctate laterally) 6

6. Lateral contour of temple slightly convex in dorsal view (Fig. 64); central tergites longer than broad; tibia 3 pale testaceous, whitish at base only, and with sub-basal and apical dark annules (frons closely punctate laterally) .... cf: **paludosa** (p. 35)

– Lateral contour of temple approximately straight in dorsal view (see Fig. 65); central tergites subquadrate; *either* tibia 3 whitish, both dorsally and at base, conspicuously darkened ventrally, and with sub-basal and apical dark annules – *or* frons sparsely punctate laterally 7

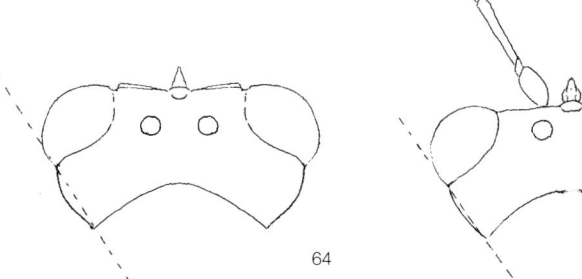

64   65

7. Frontal punctures strong, many about same diameter as interspaces; lower temporo-genal region with punctures stronger and closer (see Fig. 66); tibia 3 whitish dorsally and at base, conspicuously darkened ventrally (with sub-basal and apical dark annules); tubercles above antennal sockets stronger; costulae usually more or less erased in females (frontal horn with a narrow ventral keel – neither convoluted or swollen (cf. Fig. 58); mandibular flange wider). Males with mandibles yellow-marked, metasoma entirely blackish. *Widespread* .... **extincta** Ratzeburg (p. 37)

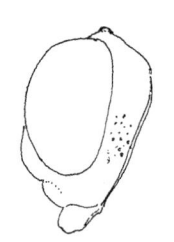

66

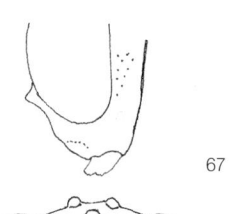

67

– Frontal punctures weaker, relatively scattered (narrower than interspaces) (Fig. 68); temporo-genal region with scattered, small punctures (cf. Fig. 67); tibia 3 testaceous, white at base, with sub-basal and apical dark annules; tubercles above antennal sockets only weakly convex; costulae usually well-developed in both sexes (frontal horn often of complex form (Fig. 70), mandibular flange narrower). Males with mandible piceous, metasoma often with reddish incisures (sometimes more extensively reddish) 8

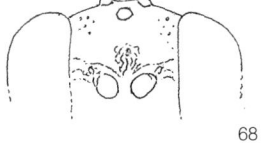

68

8. Frontal horn blade-like (as. Fig. 69), with an inverted V or U-shaped dorsal cavity, the ventral keel straight; sometimes with a more or less elliptic ventral cavity; lateral protuberances of horn small, the keels and cavities neither convoluted nor asymmetric. *Common* .... *lapponica* Holmgren (p. 37)

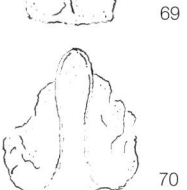

69

– Frontal horn usually with a keyhole-shaped dorsal cavity and prominent lateral protuberances (Fig. 70) – these structures to a greater or lesser extent enlarged and convoluted – often asymmetric – the ventral cavity often 8-shaped. *Frequent* .... *ceratites* Grav. (p. 38)

70

Species accounts for subgenus *Conoblasta*

Glypta paludosa sp. nov.

Female: Flagellum 1 length/breadth ratio: 5-5.5. Temples convex in dorsal aspect, laterally biplanar with respect to gena. Horn: ventral keel moderately convoluted. Interocellar space $\pm$ > than posterior ocellus to eye, and 0.75 to nearly equal to distance to occipital carina. Maximum temple length at least 0.8 x flagellar segment 1; minimum genal length $\pm$ > than 0.66 x malar space. Malar space 0.8-1.2 x basal width of mandible. Fore wing length 5-5.5 mm. Speculum reaching 0.3 to 0.4. x distance to epicnemium. Propodeum: areation, from no dorsal carinae to almost complete, inclusive of areae externae, but not dentiparae. Metasoma: Tergite 1: length/breadth ratio about 1.5-1.6; sculpture: grooves strongly acute; dorsal carinae to about 0.66 x length of tergite. Tergite 2: distinctly longer than broad, sculpture as first tergite. Ovipositor length up to combined lengths of metasoma plus propodeum. Colour: head, thorax and metasoma piceous; legs: hind tibia testaceous, base white, dark apical and sub-basal annules.

Male: agreeing in all but primary sexual characters.

HOLOTYPE. ENGLAND: female, Cambridgeshire, Chippenham Fen, Malaise trap, carr at reedbed edge, 20-30.vii.1985 (J. Field) [NMS].

PARATYPES. ENGLAND: 4 females, two males: Cambridgeshire, Chippenham Fen, Malaise trap, carr at reedbed edge, 25.vi.- 5.vii.; two females: 6-20.vii.; two females, 20-30.vii.1985; one female, Devon, Plympton, ex *E. nigricana* in *Stachys*, 27.iv.96, em. 20.v.96 [NHM]; 6 females, Surrey, Ashtead Common, Malaise trap, willow scrub, 15.vi.-6.vii., 6-20.vii., vi-22.vi.1994 [HM]; one male, one female, '*E. nigricana*, Mr. South', Billups Coll.; one male, one female, 'ex larva of *nigricostana*, from stem of *Stachys*', May 1934, 3.vi.34', Morley Coll. [NHM]. **SCOTLAND**: one female, Sutherland, Achfarg [NMS].

Taxonomy: This species has been confused with *G. elongata* in the past (sometimes also with the *ceratites* aggregate). The form of the head is more or less intermediate between that of *elongata* and that of the *ceratites* complex, and the differential is best appreciated in the presence of adequate comparative material. The name '*paludosa*' refers to the swampy habitat frequented by the species.

Distribution, abundance and phenology: Widespread and not uncommon in the UK. Flight period: late June through July.

Biology: Host: *Endothenia nigricostana* (Haw.), see above.

Glypta elongata Holmgren, 1860

Taxonomy: The conspicuously swollen temples render *G. elongata* distinctive within *Conoblasta*. However, the boundary between this and the next species is a subtle one, so far as some variants are concerned.

Distribution, abundance and phenology: Widespread and often common in fens and other wetland localities in the UK, including parts of Scotland – at least as far north as Wester Ross, Gareloch. Ireland (O'Connor *et al.*, 2007): Armagh; Down; Donegal. Flight period: July to September.

Biology: The biotope is marshland and fen. Morley (1908) gives *Bactra lancealana* (Hueb.) as host, from Fletcher. I have not personally encountered this material, but the cited host species is common and widely distributed in the biotope in which *G. elongata* occurs.

Glypta monoceros Gravenhorst, 1829

Taxonomy: *G. monoceros* can usually be distinguished from *elongata* on the basis of colour alone, but this criterion is by no means entirely reliable (see below). For this reason, proper attention must be given to structural differences in the morphology of the head.

Distribution, abundance and phenology: Much less common than *elongata*, although it has a wide distribution in England, extending also to Ireland, and northwards at least to south west Scotland.

England: Suffolk, Barton Mills, Brandon Road Heath; Middlesex, Monksmere, Hendon; Cambridgeshire, Wicken Fen; IOW, Carlsbrooke [NHM]. One male, Norfolk, Sutton High Fen, ex *?Clepsis spectrana* (Treits.), *Myrica*. 9.vi. [NMS]; Norfolk, several localities [HM]. Yorkshire, Lindrick GC, Nickerwood Ponds, Shipley Glen, Askham Bog [WE]. Scotland: a few entirely melanistic specimens from Dunbartonshire, Possil Marsh [UM]; Ireland: records extend from Cork to Donegal (O'Connor *et al.*, 2007), also Curracloe and Glenasmole (Stelfox) [NHM]. Flight period: July and August.

Biology: Found in similar wetlands habitats to the previous species. There is a reputed rearing from *Clepsis spectrana* (Treits.) on *Myrica* (Morley – f. Bridgman / Bignell), see also above.

Glypta woerzi (Hedwig, 1952)

Taxonomy: The large size and distinctive frontal horn structure form good recognition characteristics for the species.

Distribution, abundance and phenology: *G. woerzi* is little known in the UK – although probably overlooked (it should be noted that none have been taken other than through rearing from the host).

England: there are two specimens amongst material in the NHM collection, both taken prior to the original description of the species: England: West Suffolk, 1925 (Harwood); Cambridge (ex *gentianaeana*), 1949 (Pelham-Clinton). More recently: one female, Cambridgeshire, Block Fen, ex *Dipsacus*, host coll. x.2009, (JPB). Phenology unknown – no wild caught data for the adult stage.

Biology: The host is *Endothenia gentianaeana* (Huebn.) in teasel heads. Habitats include meadows, commons and roadside verges, where the host food plant grows. Bearing in mind the fact that the author reared *G. woerzi* from its host after only four attempts, it seems quite likely that the species is simply non-collectable other than via host-rearing.

Glypta fronticornis Gravenhorst, 1829

Taxonomy: Typical examples of *G. fronticornis* are fairly easy to recognise on the basis of colour patterning of the metasoma. However, structural differentiation from closely related species is weak, and pigmentation is not entirely reliable (particularly in males).

Distribution, abundance and phenology: Of uncertain distribution, commoner in the north. England: Staffordshire, Cannock Chase; Shropshire, Crowsnest Dingle, on *Olethreutes mygindiana*, *Vaccinium*; Derbyshire, Beeley Moor, ex tortricid larva on *Vaccinium*. Scotland: three males: Perthshire, Schiehallion; Inverness, Aviemore; S. Uist, Bornish (ex indet. host on *Myrica*). [NMS]. In addition, there are two examples labelled 'Farham, SR, ex viburnana' in the NHM collection. Ireland: unverified records from: Armagh; Donegal; Down; Mayo (O'Connor *et al.*, 2007). Flight period: the few wild caught data indicate that the species is on the wing during July.

Biology: *G. fronticornis* has been repeatedly reared from *Olethreutes mygindiana* (D. & S.) on *Vaccinium*, implying an association with moorland, especially in upland territory. Host larvae have been collected during March.

ceratites group:

The following three species are closely similar, and study of a large number of specimens shows some degree of overlap between them. In particular, *ceratites* and *lapponica* are only separable on the basis of form of the frontal horn – which in reality exhibits every stage in the transition between the conditions which typify the two species.

Glypta extincta Ratzeburg, 1852

Taxonomy: The characters given in the key will usually serve to distinguish this species. However, there is an area of overlap with *G. ceratites* (sensu lato) in some individuals.

Distribution, abundance and phenology: Widespread and relatively common in the UK, although poorly represented in older collections. Flight period: June and July.

Biology: *G. extincta* has been reared from *Ancylis gentianaeana* on *Salix aurita* – Isle of Skye, Trotternish Ridge, coll. 9.ix. [NMS]. The species occurs in a range of habitats, including both open country and scrub.

Glypta lapponica Holmgren, 1860

Taxonomy: Distinguished from *G. ceratites* purely on the basis of the blade-like horn structure (colour characters cited in the literature are unreliable). However, intermediate states between this and the highly complex horn structure seen in 'typical' *ceratites* are not of rare occurrence.

Distribution, abundance and phenology: Widespread and common during recent decades, although poorly represented in older collections. Ireland (O'Connor *et al.*, 2007): Armagh; Donegal. Flight period: June to September.

Biology: Recorded hosts are numerous (although it should be borne in mind that *G. lapponica* is probably an aggregate species). Typical forms have been reared from *Aphelia viburnana* (D. & S.)

– also *Pandemis cerasana* (Huebn.) on *Lonicera* – indet. 'micro' species on *Myrica*, Juniper. Examples with modified horn (truncate or spatulate) have also been reared from *Ditula angustiorana* (Haw.) and from indet. 'micros' on *Lonicera*. The species has been taken in a wide range of different biotopes, including both woodland and open country.

Glypta ceratites Gravenhorst, 1829

Taxonomy: Typical examples are readily identified on account of the much enlarged, asymmetrically convoluted frontal horn. However, as stated above, intermediate states between this and the 'simple' horn of *lapponica* are encountered.

Distribution, abundance and phenology: Widespread and not uncommon. Ireland (O'Connor *et al.*, 2007): Antrim; Armagh; Donegal; Down. Flight period: July and August.

Biology: *Glypta ceratites* has been reared from a very wide range of host species in a diversity of habitats. The following list excludes imprecise host determinations such as 'tortricid' or 'micro': *Cnephasia lacunana* (D. & S.) on *Epilobium / Plantago*; *Anarsia spartiella* (Schrank) on *Genista*. Forms with horn intermediate with *lapponica*, from *Agonopterix nervosa* and *A. ulicitella* on *Genista*, also from *A. conterminella* on *Salix; Aphelia viburnana* (D. & S.) on *?Dryas*. Host plants of unidentified lepidopterans include: *Urtica, Tanacetum, Rhamnus, Lotus* and *Myrica*.

Glypta subgenus *Glypta*

Kuslitsky (1981) introduced several valuable new characters for *Glypta* s. str., which have been utilised in the present publication, particularly in the *haesitator* group.

The key does contain some degree of complexity in the form of 'usually' and 'either / or' conditions. The aim here has been to avoid misdirection in a taxonomically difficult genus, in which misdetermination is all too easily attained on the basis of an oversimplified diagnostic key. The term 'section' is used below for assemblages of species lying above the level of the usual concept of 'species group'.

Key to sections and species groups for subgenus *Glypta*

1. Genal carina not connecting to occipital carina, temporo-genal region indented at temporal junction (Fig. 71); mandible with a wide ventral flange (Fig. 72); front coxa with a deep depression (cf. Fig. 73); clypeus somewhat weakly separated from face (Fig. 74) ....
.... ***nigrina* Section (p. 40)**

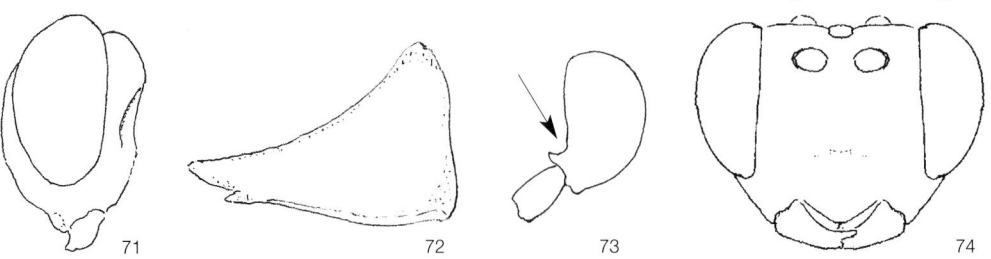

71   72   73   74

– Genal carina connecting to occipital carina, temporo-genal region with no indentation; mandible with a narrow ventral lamina; (coxa with or without depression; clypeus generally clearly separated from face) 2

2. Frons with a raised crest between antennal bases, punctate medially as well as laterally (Fig. 75), clypeus and mandibles bright yellow-marked .... ***consimilis* Section** (p. 40)

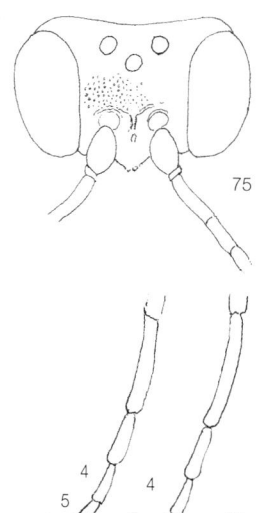

75

– Frons lacking inter-antennal tubercle – usually partly impunctate medially, (clypeus and mandibles usually not extensively yellow-marked) .... ***glypta* Section** 3

3. Segment 5 of hind tarsus often less than 1.3 x longer than segment 4 (Fig. 76a), all the tarsal segments dark testaceous (or infuscate) in colour, including their bases. *Usual hosts in pine resin galls* .... ***resinanae* group** (p. 41)

– Segment 5 of hind tarsus distinctly more than 1.3 x longer than segment 4 (Fig. 76b), (tarsal segments generally with pale bases, or entirely pale). *Not associated with conifers* 4

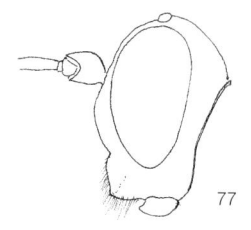

76

4. Clypeus with dense hair tuft; apex of clypeus usually 'shelved' almost at right angles (Fig. 77) – excepting rarely, when metasoma with at least tergite 2 of red or testaceous ground colour; (occipital carina usually distinctly sinuate (as Fig. 77); fore coxae often with a deep depression, or curved (Figs 78, 79)). *Hosts endophagous in stems, roots or pods, mostly of Asteraceae or Fabaceae (one rare species in roots of* Eryngium) .... ***haesitator* group** (p. 42)

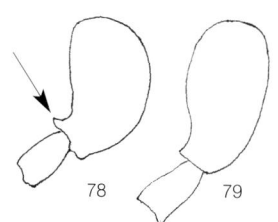

77

– Clypeus with normal pubescence, evenly convex towards apex (Fig. 80) – in case of doubt, occipital carina almost straight in profile; metasoma rarely with widespread red or testaceous colouring; fore coxae with no depression, not (or only weakly) curved 5

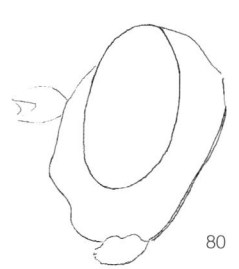

78   79

5. Occipital-genal carina almost straight in profile, the temple and gena coplanar (cf. Fig. 80), malar space usually greater than mandible width; maximum temple length usually not or only a little shorter than dorsal length of flagellum 1; interocellar distance greater than 0.5 x distance between ocellus and occipital carina; ovipositor longer than body *and* tergite 2 from longer than broad, to at most 1.3 x broader than long; malar space usually not less than mandible width. *Known hosts in roots of* Artemisia .... ***bifoveolata* group** (p. 47)

80

– Occipital-genal carina usually arcuate or sinuate (cf. Figs 77, 81), temporo-genal surface frequently biplanar; malar space at most equal to mandible; maximum temple length usually less than flagellum 1 (if rarely more than, then interocellar distance only about 0.5 x ocellus to occipital carina); if ovipositor longer than body, tergite 2 is 1.3 to over 1.6 x broader than long; malar space often narrower than mandible width. *Hosts in plant stems, seed heads or else arboreal* ... 6

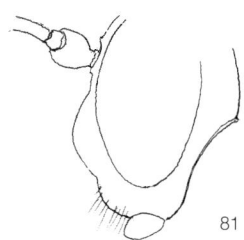

81

6. Mesopleural interstice reddish, scutellum often red (ovipositor longer than body). *Not uncommon* .... *scutellaris* group (p. 48)

–   Mesopleural interstice concolourous with adjacent cuticle (if rarely, scutellum reddish, then ovipositor much shorter than body) 7

82

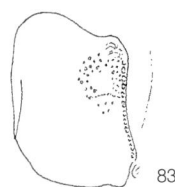 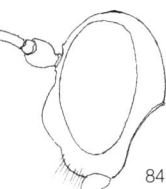  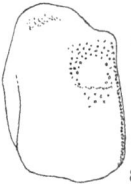 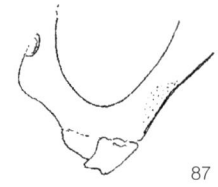

83   84   85   86   87

7. Ovipositor usually not less than body length; propodeum with the costulae usually absent or weakly developed; frons with at most a little transverse wrinkling above antennal scrobes (e.g. Fig. 82); speculum small, usually many rows of punctures behind (Fig. 83); (malar space narrower than mandible width; minimum genal length usually around 0.5 (rarely about 0.66) x width of mandible base; temporo-genal surface usually coplanar, genal carina always distinctly arcuate (cf. Fig. 84)). *Hosts in seedheads or stems of Asteraceae* .... *mensurator* group (p. 49)

–   Ovipositor usually much less than body length; costulae often strongly defined; frons with distinct arcuate striae above antennal scrobes (Fig. 85); speculum large – with few or no punctures behind (Fig. 86); (malar space usually not less than mandible width; minimum genal length usually not less than 0.66 x width of mandible base; temporo-genal surface often distinctly biplanar (Fig. 87), genal carina frequently only weakly arcuate). *Known hosts leaf-rollers on shrubs or mature trees* .... *pedata* group (p. 50)

Species account for *nigrina* section

Glypta nigrina Desvignes, 1856

Taxonomy: *G. nigrina* is easily recognised on account of the incomplete genal carina, indented temporo-genal region and short front tibial spurs (fore wing length around 5 mm).

Distribution, abundance and phenology: Uncommon, although widespread, distribution extending at least to southern Scotland. Ireland (O'Connor *et al.*, 2007): Mayo. Perhaps more frequently reared than caught. Flight period: June to August.

Biology: The habitat is deciduous woodland. There are regular confirmed rearings from: *Ptycholoma lecheana* (L.) on *Quercus*, *Tortrix viridana* (teste Lyle), *Pandemis cerasana* (Huebn.). [NHM]. The Bridgman Collection [NCM] contains rearings recorded as being from *Apotomis capreana* (Huebn.) and *Epinotia solandriana* (L.).

Species account for *consimilis* section

Glypta consimilis Holmgren, 1860 [*parvicornuta*]

Taxonomy: The sculpture of the frons renders this species relatively easy to identify.

Distribution, abundance and phenology: Widespread, although somewhat uncommon, occurring at least as far north as Inverness-shire. Ireland (O'Connor *et al.*, 2007): Dublin. Flight period: June to August.

Biology: Habitats seem to tend towards scrub on open ground such as heathland. I have encountered rearings from: *Pandemis cerasana* on *Sorbus aucuparia*, *Apotomis betuletana* (Haw.) and *Anacampsis blattariella* (Huebn.) on *Betula*; '? *Apotomis sauciana*' (Froel.) on *Vaccinium*; *Agonopterix conterminella* on *Salix*. [NMS].

Keys and species accounts for *glypta* section

Key to species for *resinanae* group

Hosts are Tortricidae in conifer galls. Both are rarely collected.

1. Occipital carina very narrowly interrupted centrally, flagellum 1 less than 5 x longer than broad, flagellum 2 nearly 4 x longer than broad (Fig. 88); maximum temple length about 1.5 x minimum genal length; tergite 1 slightly longer than broad; central tergites often greater than 1.4 x broader than long; larger (fore wing length 6-7 mm) ..... *resinanae* Hartig

88

– Occipital carina broadly interrupted centrally, flagellum 1 at least 5 x longer than broad, flagellum 2 around 3 x longer than broad (Fig. 89); maximum temple length often less than 1.5 x minimum genal length; tergite 1 not less than 1.5 x longer than broad; central tergites usually less than or equal to 1.4 x broader than long; smaller (fore wing length 5-6 mm) .... *tenuicornis* Thomson

89

Species accounts for *resinanae* group

Glypta resinanae Hartig, 1838

Taxonomy: The *resinanae-tenuicornis* species pair is not difficult to distinguish from other subgroups of *Glypta,* provided that due care is taken to accurately observe the differences in leg structure given in the key. Where overlap is apparent, other characters given in the key should suffice.

Distribution, abundance and phenology: Might prove to be co-extensive with the distribution of suitable hosts. However, existing records are too sparse to allow any broad generalisation in this respect.

England: six examples with no data (Desvignes); also: Surrey, Horsell [NHM]. More recently: one female, Surrey, Chobham, ex *Rhyacionia* on *Pinus sylvestris*, 3.iv.1992; one male, Hampshire, New Forest, Dockens Water, ex '*Blastethsia* (*Pseudococcyx*) or *Rhyacionia*', 1971 [NMS]. Flight period: at present there are no wild caught data for this species.

Biology: On the basis of the rearing data cited above (as well as through previously published information), the biotope is clearly coniferous forest (*Pinus*).

Glypta tenuicornis Thomson, 1889

Taxonomy: There should be no problem in distinguishing this species from *resinanae*.

Distribution, abundance and phenology: As with *resinanae*, the species should be looked for wherever suitable hosts are to be found. However, existing records are sparse and include no recent captures.

England: four females (Capron, Morley coll.); five females (Capron); one female, West Suffolk (Nurse); one female, Cambridge (Lyle) [NHM]. Phenology unknown.

Biology: As with the previous species, *G. tenuicornis* is found in coniferous forest. It has been reared from *Cydia grunetiana* on *Larix* (Denmark) [NMS].

Key to species for *haesitator* group

G. rufata and *G. lineata* are placed here purely as a matter of convenience. The remaining species are linked to hosts feeding in roots, stems or fruiting bodies, and probably constitute a natural species group. Characters of the clypeus, front coxae and mandibular sculpture can be difficult to appreciate, thus need special care with observation.

1. Metasoma entirely reddish testaceous; clypeus less densely hairy, not 'shelved' at apex (Fig. 90); occipital carina broadly interrupted in female; (coxa 1 unmodified, fore wing length about 3.5 mm). *Rare* .... *rufata* Bridgman (p. 44)

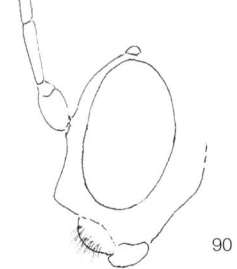

– Metasoma never entirely red; clypeus usually densely hairy, and generally 'shelved'; occipital carina at most narrowly interrupted; (coxa 1 often noticeably curved, sometimes with a frontal depression) 2

2. Thorax and metasoma testaceous with darker markings; clypeus evenly convex, antennal scapes normally cylindrical; (wing length about 3.5 mm). *Rare* .... *lineata* Desvignes (p. 44)

– Thorax black, metasoma black or partly red, or reddish marked; clypeus 'shelved', or antennal scapes flattened dorsally 3

3. Antennal scape flattened dorsally (Fig. 91); clypeus evenly convex and less densely pubescent (Fig. 92); metasomal tergite 1 up to 1.5 x longer than broad (Fig. 93); (speculum small – reaching much less than half way to epicnemium from posterior edge of mesopleurum (Fig. 94); tergites 1-2 conspicuously red-marked, at least on basal and apical margins; fore wing length 3.5-5.5 mm). *Little known in the UK* .... *ulbrichti* Habermehl (p. 44)

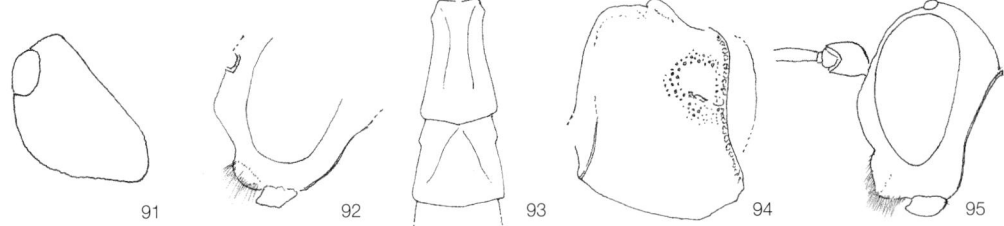

– Antennal scape with convex dorsal surface; clypeus conspicuously 'shelved' and very densely pubescent (Fig. 95); tergite 1 at most 1.5 x longer than broad; (speculum often larger – reaching half way to epicnemium; tergites 1-2 usually predominantly black in common species) 4

4. Fore coxa distinctly curved in lateral view, or with a deep depression (Figs 96, 97); (coxae usually at least basally blackish; fore wing length usually greater than 5 mm) 5

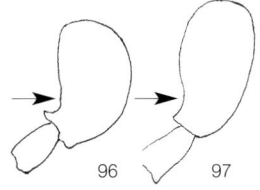

– Fore coxa not or scarcely curved in lateral view; (coxae often entirely red) 6

5.   Fore coxa curved, with a deep U-shaped depression near apex at position of bend (Fig. 96), the coxae often only darkened towards base; metasoma with central tergites generally extensively reddish; maximum temple length a little less than 2 x minimum genal length in females; ovipositor not longer than fore wing, and approximately equal to metasoma plus propodeum. *Rather rare* .... ***femorator*** Desvignes (p. 45)

—   Fore coxa smoothly curved towards apex (Fig. 97), entirely black; metasoma predominantly piceous; coxae usually predominantly piceous; maximum temple length distinctly greater than 2 x minimum genal length in females; ovipositor about 1.3 x fore wing, much longer than metasoma plus propodeum. *Not uncommon* .... ***vulnerator*** Grav. (p. 45)

6.   Mandible with numerous distinct punctures (Fig. 98); (trochanter 3 usually entirely red, ovipositor sometimes longer than front wing). *Hosts associated with Apiaceae or Asteraceae* 7

98

—   Mandible appearing almost impunctate, at most with scattered, superficial punctures (Fig. 99); (trochanter 3 often at least partly darkened, ovipositor usually shorter than front wing). *Hosts often in Fabaceae pods* 9

99

7.   Central tergites entirely red in females; (fore wing length not less than 8 mm, maximum temple length 2 x malar space or a little less). *Rare* .... ***longispinis*** (Gmelin) [*rubicunda*] (p. 45)

—   Central tergites predominantly or entirely black in both sexes; (fore wing length 4.5-7.5 mm). *Known hosts associated with Asteraceae* 8

8.   Ovipositor longer than front wing, and subequal to body length; minimum genal length less than interocellar distance; maximum temple length greater than 2 x minimum genal length; flagellum 2 less than 2 x longer than broad; malar space not narrower than width of mandible base; metasoma more slender – tergite 1 not less than 1.5 x longer than broad; speculum small – reaching much less than half way from rear edge of mesopleurum to epicnemium. *Rare* .... ***nigricornis*** Thomson (p. 46)

—   Ovipositor not longer than front wing; minimum genal length at least equal to interocellar distance; maximum temple length less than 2 x minimum genal length; flagellum 2 at least 2 x as long than as broad; malar space less than width of mandible base; metasoma less slender – tergite 1 much less than 1.5 x longer than broad; speculum larger – reaching about half way to epicnemium. *Common* .... ***similis*** Bridgman (p. 46)

9.   Frons uniformly punctate (Fig. 100); coxae predominantly black, minimum genal length at most 0.5 x maximum temple length; (malar space about 0.66 to about 0.8 x mandible base; hind trochanters red). *Widespread* .... ***haesitator*** Grav. (p. 46)

100

—   Frons with distinct transverse sculpturing in part (Fig. 101); coxae predominantly red; minimum genal length at least 0.5 x maximum temple length; (malar space 0.75 to somewhat wider than mandible base; hind trochanters usually darkened). *Abundant* .... ***trochanterata*** Bridgman (p. 46)

101

43

Species accounts for *haesitator* group

Glypta rufata Bridgman, 1887

Taxonomy: This is a very distinctive species (although somewhat 'forced' into the present species group). There is little sexual dimorphism in *rufata*, thus males should be easily recognisable from characters given in the key.

Distribution, abundance and phenology: Rare – possibly restricted to fenland.

England: Bridgman types from Cambridgeshire, Wicken Fen [NCM]; same locality, (coll. Kerrich); Essex, Colchester (Harwood) [NHM]. Recent material from: Norfolk, Sharp Street, Catfield, 11-14.vii.1983; Great Fen, Catfield, 13-14.vii.1983; male, Norfolk, Barton Turf, *Phragmites-Cladium* fen, 11-6.vi.1984 [NMS].

Biology: Most probably a species of ancient fenland. Reared from: '*Eupoecilia notulana*', Cambridgeshire, Wicken Fen (Bridgman type).

Glypta lineata Desvignes, 1856

Taxonomy: *G. lineata* is readily determined on the basis of its distinctive colour pattern.

Distribution, abundance and phenology: Little known in the UK.

England: 'Historical': Desvignes type material (no data); Suffolk, Lowestoft Dunes, beaten from *Populus* (Morley) [NHM]. More recently: Cambridgeshire, Chippenham Fen, Malaise trap; also: Kent and Gloucestershire, Dymock [NMS]. Flight period: trap catches show a late flight period, covering late August into September.

Biology: This species has been reared from *Gypsonoma aceriana* (Dup.) [NMS], and possibly belongs to that category of parasitoid which is best searched for by rearing potential host larvae. Habitats include both fenland and coastal locations, with an obvious link to *Populus* (food plant of the known host).

Glypta ulbrichti Habermehl, 1926

Taxonomy: *G. ulbrichti* does not fit into any species group with complete clarity. However, it is highly distinctive on account of the unusual form of the antennal scapes in the female sex. There is considerable variation, both in colour and in size. The type, plus one female amongst the group of specimens detailed below, has the central metasomal tergites predominantly reddish in colour, whereas other examples are more or less piceous, as well as being of distinctly smaller size.

Distribution, abundance and phenology: Known in the UK from England: two females, three males, Surrey, Leatherhead, ex *Commophila aeneana* (Huebn.) on *Senecio jacobaea* (P. Sterling) [JPB]. Flight period: unknown.

Biology: The host of *Glypta ulbrichti* is of rare and local occurence, inhabiting meadows, commons and other open territory.

Glypta femorator Desvignes 1856

Taxonomy: As already indicated, appreciation of the coxal characters of some members of this species group can be difficult. Colouration provides some measure of support regarding *G. femorator*, but cannot be relied upon in the absence of anatomical data.

Distribution, abundance and phenology: Rather rare, with scattered records extending northwards to Midlothian. It is difficult to ascertain whether or not this is merely an overlooked species.

England: Holotype male, 'Gt. Britain'; Suffolk, Stanstead Wood (Morley); Devon, Bovey Tracey; two females, 'Deal 1856' [NHM]; Sussex, Robertsbridge; Oxfordshire, Yarnton Mead; also Yorkshire [NMS]. Further examples from: Middlesex; Surrey, [HM]; Scotland: Midlothian, Gorebridge, ex *Dichrorampha senectana* (Guen.), coll. 15.i.1983, emerged: 9.v.1983 (K. Bland) [NMS]. Ireland (O'Connor *et al.*, 2007): Down. Flight period: from June into August.

Biology: The only authenticated rearing record is that given above. The known host lives on herbaceous plants growing in open territory.

Glypta vulnerator Gravenhorst, 1829

Taxonomy: Not difficult to distinguish from *femorator*. However, curvature of the coxae is a subtle character – also one prone to some degree of variation. While coxal colour is usually more or less reliable in the context of the present species pair – care must be taken not to confuse some variants with *G. haesitator*. Apart from the greater size, *vulnerator* has the ovipositor at least equal to wing length.

Distribution, abundance and phenology: Uncommon – perhaps overlooked. Existing data do indicate a wide distribution throughout England, extending northwards to Ayrshire in Scotland. Flight period: late July, into August.

Biology: I have encountered no authenticated rearing records. The species inhabits open territory, including biotopes as diverse as fen and urban parkland. A Bridgman Collection host record from *Epiblema (Eucosma) hohenwartiana* may ultimately prove valid [NCM]. There is an example reared from *Centaurea* flower head [NHM].

Glypta longispinis (Gmelin, 1790) [*rubicunda*]

Taxonomy: A very distinctive species – of which the nomenclatural history makes painful reading (and need not be repeated here).

Distribution, abundance and phenology: The species is little known in the UK.

Bridgman types: lectotype female, paratypes, two females, four males, 'Gt. Britain'. [NCM]. Ten additional specimens, all reared, England: none carrying further data, other than one labelled 'Colchester' (Harwood, Billups) [NHM]. Flight period: no wild caught data.

Biology: Originally reared from *Aethes margaritana* (Haw.) in roots of Sea Holly (*Eryngium*, Apiaceae) – thus clearly linked to coastal localities where this plant occurs.

Glypta nigricornis Thomson, 1889

Taxonomy: *G. nigricornis* is not difficult to distinguish from the other member of the species pair.

Distribution, abundance and phenology: Rather rare, although apparently quite widely distributed in Britain – at least to Midlothian in Scotland.

England: Gloucestershire; Oxfordshire; Surrey; Kent, Folkestone; Yorkshire. Scotland: Midlothian, Dalry Park [NMS]. Ireland (O'Connor *et al.*, 2007): Louth; Wicklow. Flight period: from late May, into August.

Biology: *G. nigricornis* is apparently a species of open territory. Host relationships are unknown.

Glypta similis Bridgman, 1886

Taxonomy: *G. similis* is fairly easy to distinguish on the basis of characters given here. Apart from the excessively rare *G. longispinis*, it is the largest species in the group, with wing length: 7-7.5 mm.

Distribution, abundance and phenology: Common and generally distributed throughout the UK. Ireland (O'Connor *et al.*, 2007): Down (with query). The species is quite readily reared by collecting the host pabulum during winter. Flight period: adults appear during June and July.

Biology: Hosts develop in thistle stems. Morley gives: *Epiblema scutulana* (D. & S.), in *Carduus / Cirsium* stem and roots (f. Bridgman / Fletcher). Habitats are any locality in which the hosts' food plant is found. There are confirmed rearing records from *Epiblema cirsiana* (Zell.) [NMS, NHM, HM].

Glypta haesitator Gravenhorst, 1829

Taxonomy: Care must be taken to ensure that the mandibular sculpturing is observed carefully, in order to separate this and the next species from the general milieu of *haesitator* group species. Confusion with *G. vulnerator* can be resolved by the smaller size – plus ovipositor length being at most equal to wing length in the present species.

Distribution, abundance and phenology: Probably concurrent with host distribution, although seldom collected other than by host-rearing. Records extend to south Scotland.

England: S. Devon, Dartmouth; Cambridgeshire, Gt. Shelford; Essex, Harlow. [NHM] Scotland: Edinburgh, Blackford, ex tortricid in pea-pod (*Pisum sativum*), coll. 11.viii; [NMS]. Ireland (O'Connor *et al.*, 2007): Armagh (perhaps requiring confirmation regarding determination). Flight period: occurs during July-August, into early September.

Biology: Host: *G. haesitator* has been reared from *Cydia nigricana* (F.) in *Vicia* pods, also '*Epiblema (Notocelia) rosaecolana* (Doubl.) on *Rosa*' [NMS]. Morley gives *Spilonota ocellana* (D. & S.), Roseaceae [f. Bridgman / Bignell]. Existing host data would imply a link to garden environments. The species has been introduced to North America, in the context of pest control (Dasch, 1988).

Glypta trochanterata Bridgman, 1886 Plate 4

Taxonomy: The colour of the hind trochanters generally provides an easy recognition character for a majority of examples of this species, although this is prone to some element of variation.

Distribution, abundance and phenology: Abundant and very widely distributed in the UK. Ireland (O'Connor *et al.*, 2007): Armagh; Louth. Flight period: late May to July-August.

Biology: Host: *Cydia succedana* (*ulicetana* (Haw.)) in *Ulex* pods. Habitats include heaths, commons, downland, and any other territory in which the host food plant occurs.

Key to species for *bifoveolata* group

Species in this group tend to have the density of clypeal setae almost intermediate between the *haesitator* group and other sections of the genus. However, they are quite easily distinguished on additional characters given in the key.

The *bifoveolata* group contains species associated with root-feeding tortricid larvae attacking herbaceous plants. Adults feed on umbellifer flowers.

1. Temples strongly swollen in dorsal aspect (Fig. 102); maximum temple length from a little less than, to distinctly greater than length of flagellum 1; minimum genal length distinctly greater than malar space, and wider than length of flagellum 2. Fore wing length 10-12 mm. *Rare* .... *sculpturata* Grav (below)

– Temples not, or only moderately swollen (Fig. 103); maximum temple length slightly shorter than flagellum 1; minimum genal length at most a little wider than malar space (Fig. 104), and from shorter, to somewhat longer than flagellum 2; (fore wing length often less than 10 mm) 2

2. Smaller species (fore wing length up to about 6 mm); ovipositor much longer than body length; facio-clypeal puncturation normal (Fig. 105); minimum genal length greater than malar space (Fig. 104); temples more rounded (Fig. 103). *Common* .... *bifoveolata* Grav. (p. 48)

– Larger species (fore wing length not less than 7 mm); ovipositor at most equal to body length; punctures running to striae at facio-clypeal junction (Fig. 106); minimum genal length at most equal to malar space; dorsal contour of temples nearly straight (Fig. 107). *Rare* ..... *incisa* Grav. (p. 48)

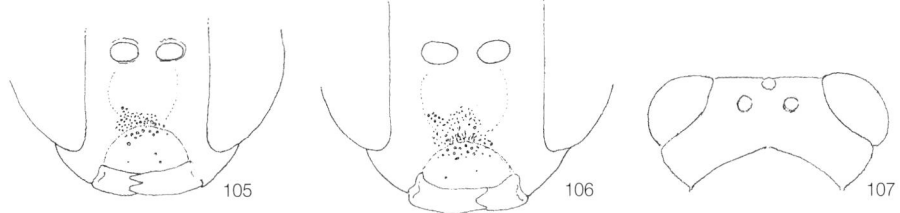

Species accounts for *bifoveolata* group

Glypta sculpturata Gravenhorst, 1829

Taxonomy: The swollen temples render *sculpturata* fairly distinctive within the *bifoveolata* group.

Distribution, abundance and phenology: Rare, although perhaps overlooked. In the UK, existing records are confined to the southern counties. Most records belong in the 'historical' realm.

England: Devon, Chudleigh; Cambridgeshire, Wicken; Kent, Tunbridge Wells; W. Suffolk, Timworth. [NHM]; Surrey, Headley Warren, Leatherhead, ix.2000 [HM]. One example coll. Morley (no data); two, coll. Cameron (no data) [NMS]. Ireland (O'Connor *et al.*, 2007): Down; Louth (unconfirmed). Flight period: September, from 'fragmentary' data.

Biology: The species has been reared from *Agapeta zoegana* (L.) in *Centaurea* roots. [NHM]. The few British records link the species to open territory, including fenland.

Glypta bifoveolata Gravenhorst, 1829

Taxonomy: The slender appearance of *bifoveolata* renders the species distinctive amongst the commoner *Glypta* species found on umbelliferous flower heads. Kuslitsky (1981) views *G. bifoveolata* as constituting a species group (so far as the European part of the former USSR is concerned).

Distribution, abundance and phenology: Very common and widely distributed, records extending northwards to Cambleton, Jura, Aberdeenshire. Ireland (O'Connor *et al.*, 2007): Armagh; Donegal; Down; Louth. Flight period: July and August.

Biology: Host: Morley (1908) gives several reputed rearings, of which *Dichrorampha simpliciana* (Haw.), a host species living in *Artemisia vulgaris* lower stem / rootstock seems the most probable (however, see data below for *G. incisa*). The biotope is open territory of any kind (obviously linked to the presence or absence of the host's food plant).

Glypta incisa Gravenhorst, 1829

Taxonomy: The largest member of the present group. The clypeal sculpture of *G. incisa* is diagnostic – although this characteristic needs careful observation.

Distribution, abundance and phenology: Rare occurrence. While existing records may appear to centre upon East Anglia, a rearing from Conwy, Wales, and one from Hertfordshire, show that the species occurs more widely.

England: Cambridgeshire, Wicken; Suffolk, Lowestoft, Ipswich [NHM]; Yorkshire: three localities (W.E.); male, two females, Hertfordshire, Welwyn, ex *Eucosma (Epiblema) foenella* (L.) [NMS]. Wales: Cardigan [NHM]; female, reared from host larvae in *Artemisia vulgaris* roots – '*Epiblema foenella* or *Dichr. simpliciana*' (Haw.); also: Llandudno Junction [NMS]. Flight period: has been taken during August, into September.

Biology: The pabulum of the host species is similar to that of the common *G. bifoveolata*.

Species accounts for *scutellaris* group

Glypta scutellaris Thomson, 1889

Taxonomy: Easily recognised from the reddish mesopleural interstice (very often supplemented by presence of a red scutellum).

Distribution, abundance and phenology: Widespread, although seldom taken in any numbers. Northwards at least to Stirlingshire, Flanders Moss [HM]. Flight period: July and August.

Biology: There are no authenticated rearing records. Habitats seem usually to be scrub and woodland.

Key to species for *mensurator* group

Species in this group attack tortricids in seed-heads of Asteraceae. Adults may be found on umbellifer flower tables, and two of the species are amongst our commonest ichneumonids. *G. rufata* (*haesitator* group) might key out here, but is easily distinguished on colour (see p. 42).

1. Tergite 1 over 1.5 x longer than broad, females with antennal scapes flattened dorsally .... cf. ***ulbrichti****, haesitator* group (p. 44)

– Tergite 1 much less than 1.5 x longer than broad, antennal scapes normal 2

2. Claws with no pectin (Fig. 108); flagellum 1 only about 3.5 x longer than broad; minimum genal length much less than width of mandible base; (fore wing length at most 6 mm). *Rare ....* ***microcera*** Thomson (p. 49)

– Claws pectinate (Fig. 109); flagellum 1 often well over 3.5 x longer than broad; minimum genal length approximately equal to width of mandible base; (fore wing length 6-9 mm). *Common and widespread species* 3

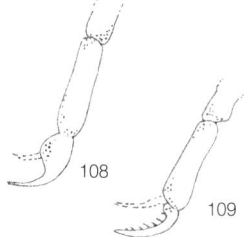

3. Interocellar distance usually greater than ocellus to occipital carina (Fig. 110); ovipositor up to 1.4 x body length or more;(central tergites often with conspicuous red markings, especially on segmental margins – sometimes entirely red). Larger on average (fore wing length 7-9 mm) .... ***nigrotrochanterator*** Strobl [*mensurator*] (p. 50)

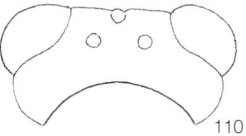

– Interocellar distance usually not greater than ocellus to occipital carina (Fig. 111); ovipositor usually not greater than 1.3 x body length; (central tergites usually with pale markings inconspicuous or absent). Smaller on average (fore wing length 6-8 mm) .... ***mensurato****r* (F.) [*lugubrina*] (p. 50)

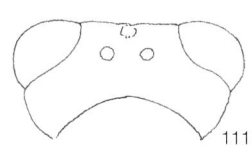

Species accounts for *mensurator* group

Glypta microcera Thomson, 1889

Taxonomy: *G. microcera* is not difficult to recognise, provided that care is taken to examine the tarsal claws carefully.

Distribution, abundance and phenology: Little known in the UK.

England: one female, Essex, Colchester – plus fifteen females, with no locality data (all coll. Harwood) [NHM]; one female, Hertfordshire, Hertford, host coll. 21.viii. [NMS]. Flight period: no wild-caught data available.

Biology: The species has been reared from *Eucosma conterminana* (Herr.-Schaeff.) on *Lactuca* [NMS]. Habitats are thus open territory in general, as with other members of the species group.

Glypta nigrotrochanterator Strobl, 1902 [*mensurator*]

Taxonomy: The familiar name for this species ('*mensurator*') was based on a misidentification. In more recent times, the name *longicauda* Hartig has been adopted on the basis of original material reputedly reared from *Bupalus piniaria* – and for which the type specimen no longer exists. This name is rejected here, on the grounds that the host connection demonstrates no possible link with the present species (an opinion shared by the late Klaus Horstmann, *pers. comm.*). In addition, the overlap between *nigrotrochanterator* and *mensurator* (s. str.) is such that no clear discrimination can be made in the absence of authentic type material. While the host association could have been (and probably was) a misidentification, it seems quite unsafe to make any definite pronouncement on the identity of the Hartig species.

So far as identification is concerned, there is certainly some degree of overlap with the next species, and it is therefore advisable to examine groups of specimens collected simultaneously.

Distribution, abundance and phenology: Very common and widely distributed in the UK, with the range extending to Shetland. Ireland (O'Connor *et al.*, 2007): Armagh; Down. Flight period: often taken in numbers on umbelliferous flower heads during July and August.

Biology: Hosts: despite the abundance of this species, information on host preference is extremely limited: 'ovipositing in *Centaurea nigra*' (G. C. Varley); collected 'on Knapweed / *Centaurea* flowers'; 'ex *Aethes sp.*'. [NHM]; also: '? reared ex *Centaurea* head' [NMS]. Habitats for *G. nigrotrochanterator* are diverse open country, given the ubiquitous nature of the host's food plant.

Glypta mensurator (F. 1775) [*lugubrina*]

Taxonomy: This species had been named '*lugubrina*' in the older literature, owing to misidentification of the true *mensurator* (see above). Discrimination between this and the previous species has, in the past, been made on the basis of taxonomically weak characters. The present treatment is based on those traits which display the least degree of overlap.

Distribution, abundance and phenology: Records are co-extensive with those for the preceding species in terms of distribution, abundance and phenology. Ireland (O'Connor *et al.*, 2007): widespread. Flight period: often taken along with *nigrotrochanterator*, June-September.

Biology: Habitats are more or identical to those of the preceding species. *G. mensurator* has reputedly been reared from *Cochylis hybridella* (Huebn.) on *Picris or Crepis* (Bridgman coll., NCM). It has been collected on flowers of *Centaurea* [NHM].

The specific status of the two species just discussed requires further research, and additional details of host preferences must play a significant role in this endeavour. Discrimination between *mensurator* and *nigrotrochanterator* should ideally be supported by controlled rearing experiments.

Key to species for **pedata** group

Members of the present species group are associated with Tortricidae feeding on shrubs in open habitats (including montane regions), or on foliage of deciduous forest trees.

1. Hind femora yellow apically, all coxae and trochanters richly yellow-marked; sub-basal band of hind tibia interrupted by white dorsally; (epistoma more prominent on average; tergite 4 always broader than long, tergite 5 at least 2 x broader than long in females; fore wing length about 3.5 mm). *Mature oak forests. Rare* .... **pedata** Desvignes [*varicoxa*] (p. 53)

– Hind femora red or black apically, coxae reddish. 2

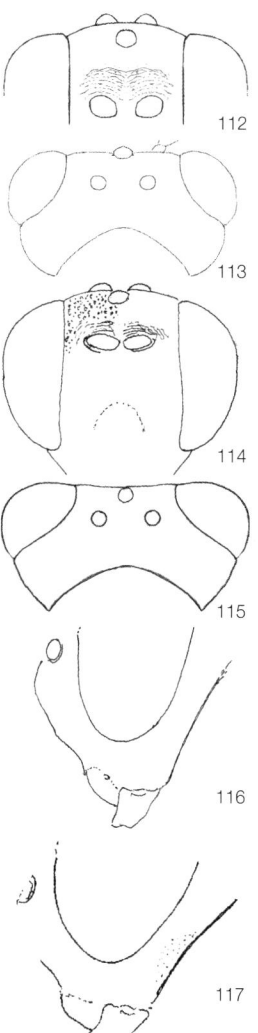

2. Frons trans-striate over its entire width (Fig. 112); temples distinctly convex in dorsal aspect (Fig. 113); central tergites at least 1.3 x broader than long; very small species (fore wing length at most 5 mm); hind tibia reddish – at most obscurely darkened at apex, and with faintly paler base; body predominantly black. *Rare, no recent records* .... **pusilla sp. nov.** [*scalaris* auctt. nec Grav.] (below)

– Frons with arcuate striae above antennal bases only (Fig. 114); temples convergent in dorsal aspect (Fig. 115); central tergites narrower (less than 1.3 x broader than long); larger species (wing length 5-8 mm); hind tibia generally darkened at apex – often white-marked over median portion; (body sometimes with varying degree of reddish suffusion – especially over basal metasomal tergites, sometimes also on scutellum) 3

3. Minimal genal length usually not exceeding 0.7 x malar space (Fig. 116); hind trochantelli reddish, the tibiae usually testaceous in centre, the sub-basal band obscure or absent (hind femur often lacking conspicuously darkened apex). *Generally associated with low shrubs, including montane regions* .... **parvicaudata** Bridgman (p. 52)

– Minimum genal length usually greater than 0.7 x malar space (Fig. 117); hind trochantelli yellowish, hind tibia usually yellowish-white centrally, the sub-basal band distinct (hind femur rarely lacking darkened blackish apex). *Hosts on mature trees* .... **pictipes** Taschenberg (p. 52)

Species accounts for *pedata* group

Glypta pusilla sp. nov. [*scalaris* auctt. nec Grav.]

Female: Flagellum with 28-30 segments; flagellum 1 about 5 x longer than broad. Temple distinctly swollen laterally in dorsal aspect. Occipital carina interrupted centrally. Frons with transverse striae over entire width. Face with punctures greater than interspaces, becoming less than, laterally. Clypeus strongly convex, punctures approximately equal to interspaces in basal half. Interocellar space about 1.3 x posterior ocellus to eye, and about 0.66 x posterior ocellus to occipital carina. Maximum temple length approximately 0.8 x flagellum; minimum genal length distinctly wider than malar space, and subequal to basal width of mandible. Fore wing length at most a little greater than 5 mm. Mesopleurum with large punctures, mostly around width of interspaces: speculum shining glabrous, reaching about half distance to epicnemium. Propodeum with complete areation, costulae emitted just before middle of area superomedia; sculpture: with large, deep punctures, mostly about width of interspaces, tending to 'run' in region laterad of subdorsal carinae. Metasoma with tergite 1 at most 1.3 x longer than broad, with dorsal carinae reaching well beyond centre, puncturation about equal to interspaces, some striation laterally. Tergite 2 nearly 2 x broader than long, with small, dense puncturation – the punctures wider than interspaces on dorsum, becoming of similar size and distribution to those of tergite 1 laterad of diagonal grooves. Laterotergite 3 less than 2.5 x longer than broad. Ovipositor: about 0.66 x length of metasoma. Colour: black – legs testaceous, hind tarsi at most faintly darkened at apex, and with obscurely paler base.

HOLOTYPE. ENGLAND: female, Thorne, 11.vii.07 (Morley) [NHM].

PARATYPES. ENGLAND: female, Devon (Marshall); female, no data (Marshall) [NHM].

Taxonomy: The name *'pusilla'* (tiny) refers to the fact that this is the smallest *Glypta* species found in the UK. It is not at all difficult to recognise on the basis of characters given herein.

Distribution, abundance and phenology: Known from a very few 'historical' specimens in the NHM and Norwich Museum collections. Ireland (O'Connor *et al.*, 2007): *scalaris auctt. nec* Grav. is erroneously recorded as a synonym of *punctifrons* Bridgman (see p. 53). The sparse wild-caught data indicate a flight period that includes July.

Biology: No authenticated rearings.

Glypta parvicaudata Bridgman, 1889

Taxonomy: This and the next species are closely similar, and it is important to pay special attention to structural (as against purely colour) characteristics in order to determine them correctly. Ovipositor length has been used to distinguish the two in the past. However, there is considerable overlap in this character.

Aubert (1978) lists *breviventris, crassitarsis*, and *tenuitarsis* of Thomson as synonyms of the present species. However, examination of material in the Thomson collection shows that this is in error.

Distribution, abundance and phenology: *G. parvicaudata* is not uncommon in suitable habitats. Records extend from Cambridgeshire and Oxfordshire to Sutherland and Jura, with more northern localities predominating. [NMS]. Ireland: Wicklow, Liffey Head. [NHM]; O'Connor *et al.*, 2007: Donegal; Mayo (unconfirmed). Flight period: most data imply a flight period during July-August.

Biology: Generally associated with low shrubs, particularly (although by no means exclusively) in more upland locations. Known hosts include *Acleris hastiana* (L.) (on *Salix repens / aurita*) / '?*Spilonota ocellana* (D. & S.) on *Myrica / ?S. atrocinerea*. [NMS].

Glypta pictipes Taschenberg, 1863

Taxonomy: Length of gena is the best structural trait for separating *pictipes* from the previous species.

Distribution, abundance and phenology: Widely distributed, although rather uncommon, range extending to northern Scotland. Ireland (O'Connor *et al.*, 2007): Laois; Wicklow (unconfirmed). Flight period: July to September.

Biology: By way of contrast to the previous species, *G. pictipes* is associated with mature forest, rather than open habitats. It has been reared from *Acleris ferrugana* (D. & S.) on *Quercus*, also *'tortricid* on *Fagus; ?Acleris* on *Betula'* .

Glypta pedata Desvignes, 1856 [*varicoxa,* syn. nov.]

Taxonomy: *G. pedata* can usually distinguished from the preceding species on the basis of colour alone.

Distribution, abundance and phenology: Of rather rare occurrence, and possibly confined to ancient oak woodland.

Two specimens with no data [NHM]; two females, one male, England: Berkshire, Windsor Forest, Malaise trap, old *Quercus, Juncus, Fagus*; one male, Berkshire, Windsor Forest, Highstanding Hill, *Quercus / Fagus / Populus tremula*; one female, *Quercus / Juncus*; one male, Surrey, Ashtead Common, old *Quercus*. [HM]; Yorkshire, Netherton. Flight period: early May to August, also September, perhaps implying two generations a year.

Biology: Hosts occur on foliage of mature deciduous trees. There are rearings from *Acleris ferrugana* (D. & S.), obtained in the oak project, Hope Department of Entomology.

On *Glypta punctifrons* Bridgman, 1889

Glypta punctifrons of Bridgman is known from the male type alone (reared from *Metendothenia atropunctana*, Perthshire: Rannoch). It was synonymised with *G. scalaris* (with query) by Aubert (1978), and has since remained in that situation. *G. punctifrons* does exhibit the extensively areolated propodeum and arcuately striate frons characteristic of the *pedata* group. However, it is certainly not conspecific with *scalaris* of Gravenhorst (nor indeed, with *scalaris* auctt. nec Grav.). Unusual traits of *punctifrons* include a deeply down-curved occipital carina, non-coplanar temple and gena, and very heavily coriaceous interspaces on the central metasomal tergites. The minimum genal length is distinctly wider than mandible base.

Additional material (especially of the female sex) of *punctifrons* will have to be obtained, before any further progress can be made with taxonomic diagnosis.

Tribe Atrophini [Lissonotini]

There should be no real difficulty in recognising the genera of this tribe following characters given in the key. As stated earlier, *Cryptopimpla* might arguably be merged with *Lissonota*: quite apart from weaknesses in characters of the female sex, males of this genus are in actual practice indefinable as a separate entity – thus best reared or collected along with females. Hosts are mostly Tortricoidea, Gelechioidea and Yponomeutoidea. Some of the larger species attack wood-boring Lepidoptera such as Sesiidae and Cossidae, and some parasitise cryptophagous Pyralidae or Noctuidae. *Cryptopimpla* species parasitise Geometrid larvae feeding in exposed positions.

Key to the genera of Atrophini

1. Areolet long-stalked (Fig. 118); genal carina meeting hypostomal at base of mandible (Fig. 119); prepectus strongly developed (Fig. 120); propodeal spiracles large and elongate – at least 3 x longer than wide (Fig. 121); propodeum with no lateral carinae *or* frons with two horn-like protuberances (Fig. 122) metasomal tergites glossy black with broad yellow markings (or if red – then frons with two horns; propodeum often with conspicuous yellow markings) ....
.... ***Syzeuctus*** (p. 55)

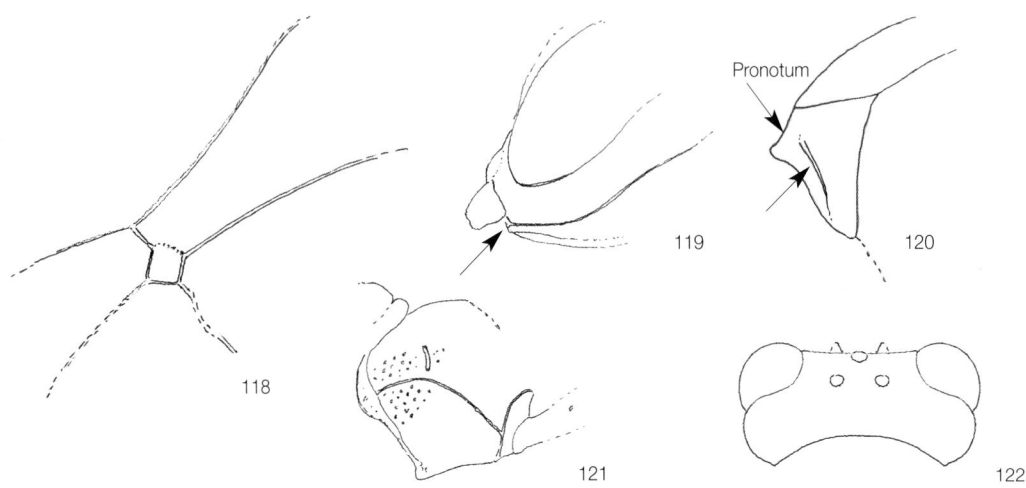

– Areolet usually short-stalked; genal carina usually meeting hypostomal behind base of mandible (Fig. 123); prepectus usually absent; propodeal spiracles from subcircular to about 2 x longer than wide – propodeum usually with lateral carinae, frons never with horns; (tergites sometimes obscurely yellowish-marked, propodeum not yellow-marked) 2

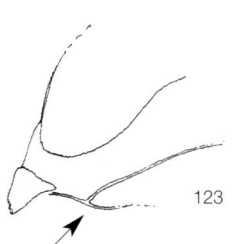

2. Head and body with long erect hairs (Fig. 124), body entirely black; central tergites distinctly broader than long (Fig. 125); ovipositor shorter than metasoma; propodeum with posterior carina absent; (fore wing length about 6 mm) .... ***Arenetra*** (p. 56)

– If rarely, the head is unusually hairy, then ovipositor longer than metasoma, and propodeum with posterior carina; (central tergites often subquadrate to longer than broad) 3

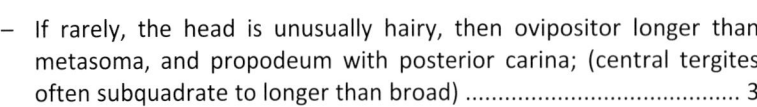

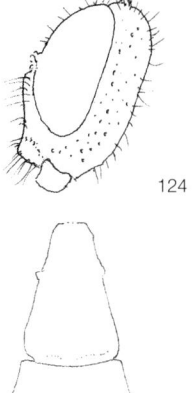

3. Nervellus intercepted at or above centre (Fig. 126); ovipositor distinctly shorter than metasoma, the sheaths strongly flattened and expanded near base (Fig. 127) – in dorsal view, about 1.5 x width of hind basitarsus; (head with longer, more dense hair (Fig. 128); hind tarsal claws distinctly pectinate (Fig. 129); fore wing length 7-8 mm) .... *Alloplasta* (p. 57)

– Nervellus intercepted below centre; ovipositor often longer than metasoma, the sheaths usually much less than 1.5 x width of hind basitarsus near base (Fig. 130); (head usually with shorter, less dense hair (Fig. 131); hind tarsal claws very often lacking distinct pectin) 4

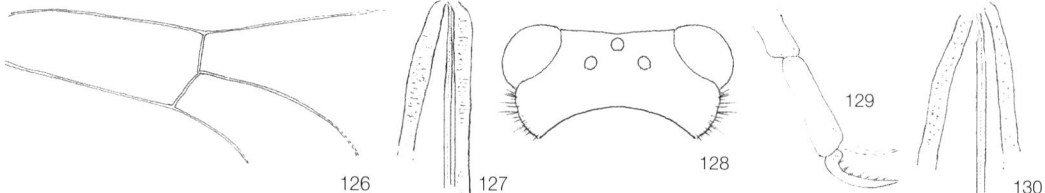

126   127   128   129   130

4. Ovipositor sheaths usually distinctly longer than 1.4 x hind tibia; subapical flagellar segments not spindle-shaped; (head most often not, or only weakly contracted behind eyes) .... *Lissonota* (p. 58)

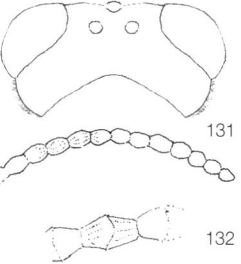

131

– Ovipositor 0.5 to 0.66 x length of hind tibia; subapical antennal segments usually more or less spindle-shaped in females (Fig. 132); (head frequently strongly contracted behind eyes) .... *Cryptopimpla* (p. 108)

132

Genus *Syzeuctus* Förster, 1869

The genus *Syzeuctus* is of worldwide distribution, occurring mostly in warm climates, and of predominantly southern distribution in Europe. The species are generally gaudily patterned. Hosts include both Pyraloidea and tortricoids. Kuslitsky (1981) treats 10 species in the European part of the former USSR.

Key to species of *Syzeuctus*

1. Frons with two horns (Fig. 133); ovipositor much longer than metasoma, latter predominantly red; propodeum with posterior transverse carina present, carina lateralis at least partly formed; fore wing without darkened areas). *Rare* .... *bicornis* (Grav.)(below)

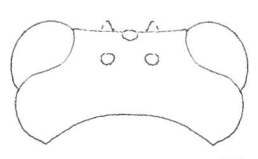

133

– Frons without horns; ovipositor around length of metasoma, latter black, with prominent yellow markings; propodeum with posterior transverse carina absent, carina lateralis absent; fore wing darkened at apex. *Uncommon* .... *fuscator* (Panzer)(p. 56)

Species accounts for *Syzeuctus*

Syzeuctus bicornis (Gravenhorst, 1829)

Taxonomy: *S. bicornis* is easily recognised by the frontal horns, in combination with the usual diagnostic features of the genus.

Distribution, abundance and phenology: The species is little known in the UK – with a single recent record.

England: Suffolk, Bury St Edmunds (Morley); Kent, Tunbridge Wells, 1920 [NHM]; Surrey, Banstead Downs, Malaise trap, 12-27.vii.2000 [HM].

Biology: There are no authenticated rearing records. The only recent capture was made on chalk downland, and potential hosts might be sought in such locations.

Syzeuctus fuscator (Panzer, 1809) [*maculatorius*] Plate 6

Taxonomy: A very distinctive species, on account of its gaudy colour pattern.

Distribution, abundance and phenology: Uncommon: apparently of predominantly southern distribution, occurring on heathland and in coastal locations.

England: Hampshire, Hayling, New Forest; Isle of Wight; Dorset, Swanage; Surrey, Box Hill; Cornwall; Devon, Plymouth; Jersey: St. Catherine [NHM]; Surrey, Thursley Common [HM]; Wales: Glamorgan, Gower [NMS]; 'Pembroke' [NCM]. Flight period: July and August.

Biology: *S. fuscator* has been reared from the phycitine pyralid *Oncocera (Pempelia) genistella* (Dup.)*,* which occurs on heaths and commons in association with *Ulex*. In addition, I have seen material reared from *P. palumbella* (D. & S.). Given that the host pabula of the aforementioned host species includes *Erica* and *Calluna*, heathland is probably the usual habitat for inland sites. There are two additional rearings from another phycitine, *Epischnia bankesiella* (Rich.), which occurs on sea cliffs. Adults have been taken on *Daucus.*

Genus *Arenetra* Holmgren, 1859

The divergent form of this genus renders it quite distinctive within the Banchinae.

– Entirely black species; ovipositor about 0.5 x fore wing length; legs unusually slender. *Widespread, early spring* .... ***pilosella*** (Grav.)

Species account for *Arenetra*

Arenetra pilosella (Gravenhorst, 1829) Plate 5

Taxonomy: This is a very clearly defined genus, unlikely to be mistaken for any other.

Distribution, abundance and phenology: Not many specimens exist in museum collections. However, this is probably at least in part due to the very early appearance of the adult stage. Existing data suggest that the species is commoner in more northern, upland regions. Flight period: Townes and Townes (1978) state that Nearctic *Arenetra* species appear during early spring or late autumn, their habitats being bare ground with sparse vegetation.

England: 'ex *pilosaria*, Lincoln' (Morley); Northumberland, Moor House NR, 20-30.iv.; Cumbria ex *Cerapteryx graminis*; Cumberland, Buttermere; 'Sussex'. Scotland: Perthshire, Ben Lawyers (Cameron); Inverness-shire, Aviemore (Harwood); Argyle, Glen Lochy, 'on snow' [NHM].

Midlothian, Peebles, Cademuir (on window); Aberdeenshire, Muir of Dinnet; Inverness-shire, Glen Moriston; Nethy Bridge, ex *Paradiarsia* (*Eugnorisima*) *glareosa* (Esp.); also ex *C. graminis* [NMS]. Dunbartonshire, Milngavie, flying low over *Calluna* [UM].

Biology: The species has reputedly (and probably erroneously) been reared from *'pilosaria'* (Morley). Authenticated records are from *Cerapteryx graminis* (L.) and *Paradiarisa* (*Eugnorisma*) *glareosa* (Esp.). Nearctic rearings of *Arenetra* are from several *Euxoa* species (Townes, *loc. cit.* above).

Genus *Alloplasta* Förster 1869

Key to species of *Alloplasta*

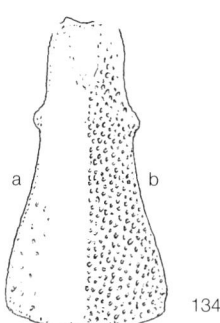

1. Metasoma predominantly red, face and mesonotum lacking yellow markings; tergite 1 with small transverse rugae, in part sparsely punctate laterally (Fig. 134a); tergite 2 quadrate. *Common and widespread* .... *piceator* (Thunberg) [= *murinus*]

\- Metasoma black, incisures pale; face and mesonotum usually yellow-marked; tergite 1 with dense, coarse puncturation (Fig. 134b); tergite 2 distinctly transverse. *Rare* .... *plantaria* (Grav.)

134

Species accounts for *Alloplasta*

Alloplasta piceator (Thunberg, 1822) [*murinus*] Plate 7

Taxonomy: The two British *Alloplasta* species differ widely from one another, thus there should be no problem in recognising the present species from the next.

Distribution, abundance and phenology: Widely distributed in Britain, and often common. Ireland (O'Connor *et al.*, 2007): widespread. Flight period: adults appear during May and June.

A. piceator has four subspecies in America (Townes and Townes, 1978).

Biology: The species frequents commons, hedgerows, woodland margins and other open or semi-open territory. Morley gives *Orthosia miniosa* (D. & S.) / *gracilis* as hosts. Recent experiments demonstrate a host preference for *O. gothica* (L.). The species will oviposit into *O. gracilis* (D. & S.), but fails to complete development therein. It does not attack *O. stabilis* (*cerasi* (F.)). [M. R. Shaw, *pers. comm.*].

Alloplasta plantaria (Gravenhorst, 1829)

Taxonomy: Easily recognisable within *Alloplasta* on the basis of colour characteristics alone.

Distribution, abundance and phenology: There are very few specimens of *A. plantaria* in UK museum collections. However, these have been collected over a fairly wide geographic area.

England: Buckinghamshire, Soulbury; Huntingdon, Woodwalton; Wye; East Suffolk; West Suffolk. Wales: Cardiff. In addition, there are a few 'historical records' lacking data [NHM]. More recent records: Cambridgeshire, Wicken Fen, 1965 [UM], Kent [NMS].

Biology: The biotope is probably similar to that of the preceding species, although ancient fens are included amongst the fragmentary data at hand. I have encountered a single authentic rearing labelled '?*Orthosia stabilis* (*cerasi* (F.))': Kent, East Blean (coll. E. S. Bradford) [NMS] .

Genus *Lissonota* Gravenhorst, 1829

Lissonota is a very large genus of worldwide distribution – although best represented in the Holarctic region. Townes and Townes (1978) treat 143 Nearctic species, with an estimated real total of over 200; eight Nearctic species are recorded as occurring also in the Palaearctic. Aubert (1978) lists around 100 species for the West Palaearctic region, whilst Kuslitsky (1981) treats 111 in the European part of the former USSR, with an estimate of not less than 150 species.

Aubert (1978) divides *Lissonota* into *Lissonota* s. str. v. *Campocineta*, also separating *Loxonota* as an additional subgenus in a later publication (1993). Aubert also places *Loxodocus* of Townes as a probable synonym of *Lissonota*. Kuslitsky's treatment of *Lissonota* (1981) is based primarily on Aubert's key.

Key to groups of *Lissonota*

The subgenera of *Lissonota* are not difficult to recognise. However, special care must be taken with regard to characters associated with the hind tarsal claws.

1. Occiput interrupted, and with a deep central depression that extends forwards almost to the line of the ocelli (Fig. 135); maximum temple length much more than 2 x minimum genal length; laterotergite 4 at least 4 x wider than third; (fore wing length about 10 mm). *Abundant in meadows* .... ***Meniscus lineolaris*** (Gmelin) (p. 60)

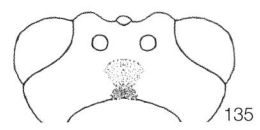

– If rarely, the occipital carina is interrupted centrally, the depression at that point is shallow and does not extend forwards to near the line of the posterior ocelli (Fig. 136); maximum temple length not greater than 2 x minimum genal length; laterotergite 4 much less than 4 x wider than third 2

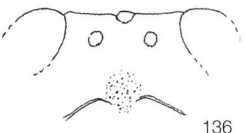

2. Last sector of radius sinuate and areolet long-stalked (Fig. 137); tarsal claws at least half pectinate; speculum absent – this area punctate (Fig. 138); mesonotum often with yellow longitudinal stripes; metasoma usually extensively red-marked. *Coastal or breckland species* .... **Loxonota** (p. 72)

– Last sector of radius evenly curved, to slightly sinuate (Fig. 139), areolet often subsessile or short-stalked – rarely absent or pentagonal 3

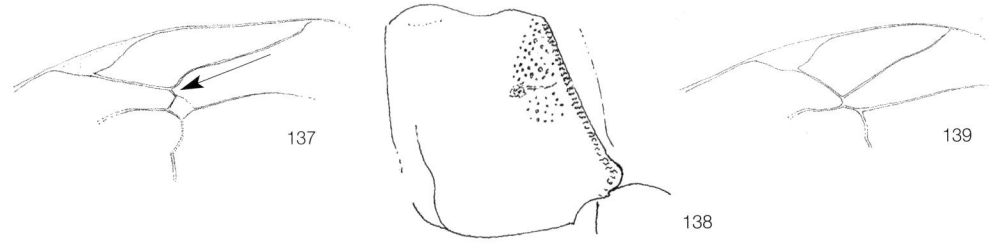

3. Tarsal claws usually at least 2 x, never shorter than length of arolium (Fig. 140) *or else pectinate to at least 0.66 x length of claw* (Fig. 141); fore wing length often much greater than 6 mm; sternites 2-4 medially dark (similar to laterosternites); mesosulcus often open, sometimes deeply excavate (Fig. 142); (minimum genal length usually more than 0.4 x length of flagellum 2 (Fig. 143)) 4

— Tarsal claws often shorter than, seldom greater than 1.5 x arolium – *at most with a few teeth towards base*; fore wing length usually in range 4-6 mm; sternites 2-4 usually medially noticeably paler than laterosternites; mesosulcus usually closed apically by one or more transverse carinae, never excavate (Fig. 144); (minimum genal length often less than 0.4 x flagellum 2) .... ***Campocineta*** (p. 75)

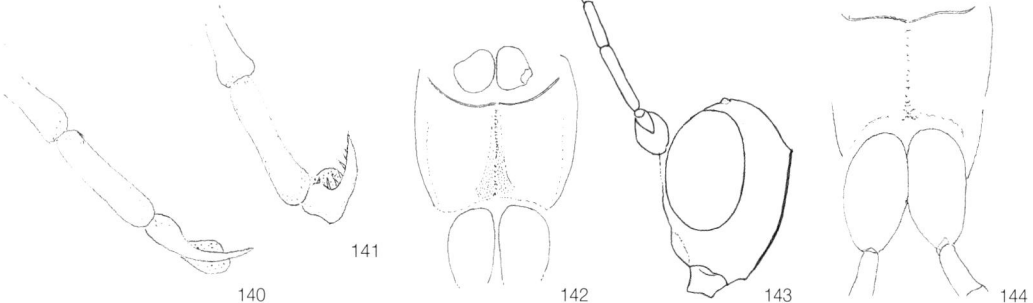

4. Metanotum around 2 x width of distance between petiolar spiracles; tergite 1 and central tergites in part with transverse microsculpture, and not extensively punctured (as Fig. 145); speculum small and dull (Fig. 146); mesosulcus *closed,* i.e. apically with one or more transverse costae strongly raised; (flagellum 1 around 5 x longer than broad; body black with red legs – hind tibia and tarsus darkened; minimum genal length around 2 x width of malar space). *Often abundant on and around rotting tree boles* .... ***biguttata*** Holmgren (p. 63)

— Metanotum usually much narrower than 2 x the distance between petiolar spiracles; if rarely, the central tergites exhibit transverse microsculpture and sparse puncturation, mesosulcus *open*; speculum usually glabrous – except when metasoma extensively red-marked; (flagellum 1 often less than 5 x longer than broad; hind tibia and tarsus usually differently coloured; minimum genal length often less than 2 x of malar space) 5

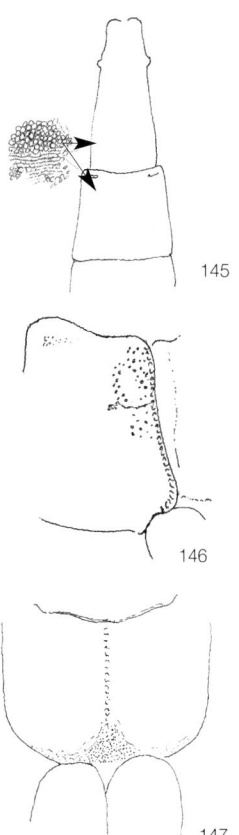

5. Mesosulcus closed posteriorly (as Fig. 144), always shallow; tarsal claws usually pectinate to over half (cf. Fig. 141). *No common species. Hosts: Cossidae and Sesiidae* .... ***Lampronota*** (p. 60)

— Mesosulcus open posteriorly, often deeply excavate (Figs 142, 147); tarsal claws often pectinate to less than half. *Several very common species, mostly attacking Noctuidae on Poaceae* .... ***Lissonota*** s. str. (p. 68)

Meniscus group

Species account for *Meniscus* group

Lissonota lineolaris (Gmelin, 1790) [*catenator*] Plate 8

Taxonomy: The head structure renders the present species very distinctive – although care should be taken to avoid confusion with the rarely encountered *Lissonota nitida*. In addition to the key characters, the interocellar distance is not much greater than ocellus to eye, and about 0.5 x the distance between ocellus and occipital carina (Fig. 135). The metasomal tergites have a dull coriaceous background sculpture.

Distribution, abundance and phenology: Widely distributed and common in the UK. Ireland (O'Connor *et al.*, 2007): Armagh; Down; Waterford. The flight period is during June and July.

Biology: Habitats are meadows, commons, and other open territory. The host is *Apamea crenata* (Huefn.). Oviposition into small larvae in flower heads of Poaceae has been observed.

Lampronota group

Key to species for *Lampronota* group

Lampronota group [= '*Meniscus*' sensu auctt.] species are seldom collected other than by rearing from their hosts (usually Sesiidae, but also including Cossidae). *L. biguttata* has been keyed separately (p. 59, p. 63).

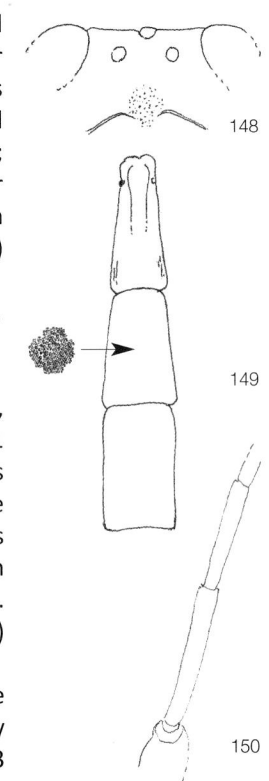

1. Occipital carina broadly interrupted centrally (Fig. 148); (minimum genal length a little greater than, to about 1.4 x malar space; interocellar distance around 2 x distance between ocellus to eye, and a little less than to somewhat greater than distance between posterior ocelli and occipital carina; metasomal tergites with shining glabrous background; ovipositor longer than metasoma. Mesonotum with yellow shoulder marks continuing to wing base, metasoma more or less suffused with testaceous colouring). *Rare* .... ***nitida*** Grav. [*agnata*] (p. 63)

148

149

–   Occipital carina not (or barely) interrupted medially 2

2. Central tergites distinctly longer than broad – uniformly coriaceous, or with a little indistinct puncturation (Fig. 149); flagellum segment 1 from 6-7 x longer than broad in females (Fig. 150); (mesosternal sulcus transversely striate throughout, frons somewhat impressed above antennal bases, raised laterally; fore wing length 9-10 mm; black, legs reddish; thorax black, sometimes with yellow lateral markings on scutellum). *Scotland, Ireland* ....
.... **plana sp. nov.** [*impressor* Grav.-Ths.] (p. 63)

150

–   Central tergites subquadrate to transverse – often strongly punctate (e.g. Figs 156, 159); (flagellum 1 at most 6 x longer than broad – usually distinctly less; mesosulcus usually only trans-costate at apex.) 3

3. Minimum genal length at least 0.4 x eye length (Fig. 151); larger species (wing length 14-16 mm); central tergites confluently punctate (Fig. 152) 4

– Minimum genal length much less than 0.4 x eye length; smaller species (wing length much less than 14 mm); central tergites with smaller, non confluent punctures (excepting rarely, when the mesonotum has a yellow shoulder stripe continuing to wing base, or the mesonotum and scutellum are yellow-marked) 5

4. Frons strongly swollen and heavily depressed above antennal bases (Fig. 153), closely punctate laterally and sparsely punctate mesally, trans-striate above antennal sockets; clypeus nearly 2 x as wide as long (Fig. 154); very large species (about 16 mm wing length); tergite 2 uniformly and rugosely punctate (Fig. 152); apical half of ovipositor strongly laterally compressed – depth up to 2 x dorsal width; minimum genal length up to 1.3 x length of flagellum 2 (Fig. 151). *On* Cossus .... ***setosa*** (Geoffroy) (p. 64)

– Frons weakly swollen laterally, weakly depressed above antennal bases, and uniformly punctate throughout – clypeus about 1.5 x wider than long (Fig. 155); somewhat smaller species (wing length about 14 mm); tergite 2 with median and posterior region band more finely punctate than rest of tergite, generally lacking rugae (Fig. 156); ovipositor less strongly compressed, less than 2 x deeper than as wide towards apex; minimum genal length well over 1.3 x length of flagellum 2. *On* Sesia .... ***fulvipes*** (Desvignes) (p. 65)

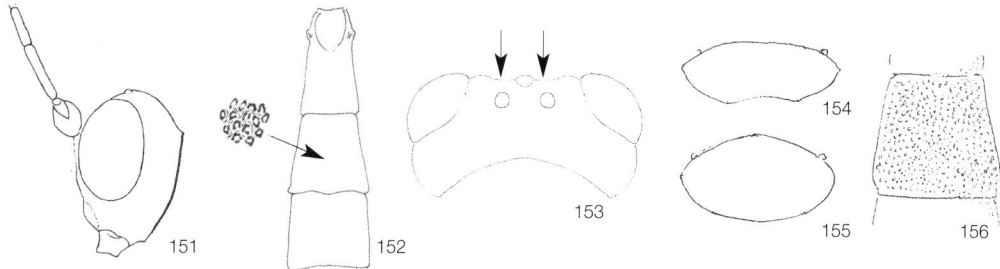

5. Mesonotum with conspicuous yellow submedian stripes; scutellum yellow-marked laterally, posterior segmental margins broadly reddish; tergite 1 entirely punctate, apart from a very narrow dorso-median impunctate band (Fig. 157); vertex sharply angled (Fig. 158) (fore wing length about 5 mm) .... ***deversor*** Grav. (p. 67)

– Mesonotum sometimes with yellow shoulder markings, but always lacking submedian fasciae; scutellum black, metasoma lacking conspicuous red areas; tergite 1 with limited areas of puncturation (if with only a median impunctate band, this is broader; vertex usually not, or only weakly angled at centre) 6

6. Tergite 1 not or but little more than 1.5 x longer than broad; tergite 2 strongly transverse, at least 1.3 x broader than long, with punctures subequal to interspaces (Fig. 159); (fore wing length around 6 mm, ovipositor < body length) 7

– Tergite 1 much more than 1.5 x longer than broad; tergite 2 subquadrate to distinctly elongate, punctures mostly narrower than interspaces; (fore wing length at least 7 mm, ovipositor often > body length) 8

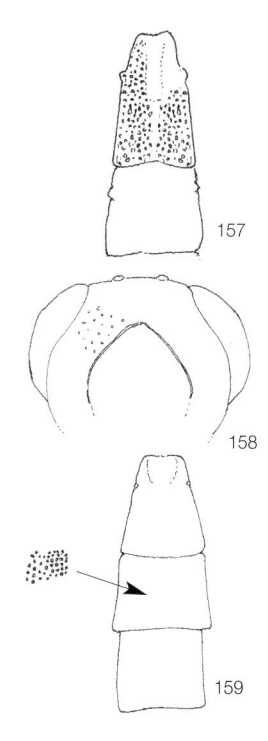

7. Flagellum 1 at least 5 x longer than broad (Fig. 160); gena narrower in relation to length of flagellum 2, and to malar space (Fig. 161); interocellar space nearly 2 x distance between posterior ocellus and eye; tergite 1 heavily and closely punctate, including a band across dorsum – the pre-apical impression in part striate (Fig. 162); yellow mesonotal shoulder marks extending to wing base, subalar prominence yellow, hind tibia testaceous – infuscate towards apex, front and middle coxae yellow-marked. *Hosts: Pyralidae on conifers* .... ***dormitor* sp. nov.** (p. 65)

– Flagellum 1 only about 2.5 x longer than broad (Fig. 163); gena broader in relation to length of flagellum 2, and to malar space (Fig. 164); interocellar space less than 1.5 x posterior ocellus to eye; tergite 1 polished, with fine evenly distributed puncturation (Fig. 165); mesonotum and subalar prominence not yellow-marked, hind tibia testaceous, whitish at base, coxae not yellow-marked. *Hosts: Sesiidae* .... ***pimplator*** (Zett.) (p. 66)

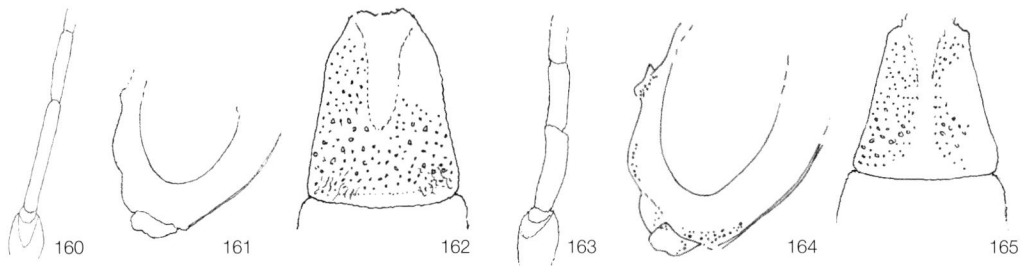

160   161   162   163   164   165

8. Mesosulcus closed by a pair of prominent tubercles (Fig. 166); (tergite 1 with only a narrow apical non-sculptured area, and with a zone of striation apically (Fig. 167)) [males: mesosternum black] .... ***freyi*** (Hellén) (p.67)

– Mesosulcus closed by weaker carinae only; (tergite 1 at most laterally striate) 9

9. Flagellum 1 from 4-5 x longer than broad; tergite 1 with a long dorso-median glabrous zone that is usually interrupted distally by a narrow transverse band of puncturation (Fig. 168); propodeum with strong dorso-median longitudinal carinae (Fig. 169), adjacent sculpture running to strong transverse rugae; hind tibia testaceous at base; (ovipositor at least equal to thorax plus metasoma); [males: tergite 1 longitudinally rugose, mesosternum black, face with yellow orbits] .... ***frontalis*** (Desvignes) (p. 67)

– Flagellum 1 from 3.5-4.5 x longer than broad (Fig. 170); tergite 1 with a larger dorsal, pre-apical obscurely sculptured area, and a wide area of puncturation across medial zone (Fig. 171); propodeum at most with weak longitudinal carinae, adjacent sculpture weakly rugose. Hind tibia basally whitish; (ovipositor shorter than thorax plus metasoma); [males: tergite 1 extensively punctate, mesosternum broadly yellow, face yellow] ....
.... ***canaliculata*** Szépligeti [*pimplator*] (p. 68)

166   167   168   169   170   171

Species accounts for *Lampronota* group

Lissonota biguttata Holmgren, 1860 [*femorator*]

Taxonomy: This species is easily confused with members of the subgenus *Campocineta*, which it superficially resembles. It will, nevertheless, key with *Lampronota*, despite being discordant within that group – both structurally and with respect to host preference.

Males differ widely from the female sex. The face is yellow-marked, the fore and middle coxae, trochanters and trochantelli are yellow. The mesonotal shoulder marks are sometimes continued backwards over the mesonotum. The central metasomal tergites have a variable degree of pale markings, especially towards the apical margins. In common with the females is the fine transverse sculpturing of tergite 1 (at least in part), plus the fuscous coloured hind tarsi (plus sometimes femora), along with the pectinate tarsal claws.

Distribution, abundance and phenology: Common and very widely distributed in the UK. Ireland (O'Connor *et al.*, 2007): Armagh; Down. There are two generations, the second of which is certainly linked to rotten timber. The first is presumed to emerge during early spring, and the second generation appears during mid-summer. Flight period: (early spring) and July-August.

Biology: *L. biguttata* is often abundant on and around rotting tree boles. Curiously, the first generation has been reared from *Operophtera brumata* (L.) on a number of occasions, yet has never been collected in the adult stage. The second generation probably attacks lepidopteran hosts in rotten wood – viz: rearings 'ex *Salix* containing *Xiphydria prolongata*' (Benson); 'from tree containing species of *Sphecoid*' [NHM]. Given the wide distribution, abundance and phenology of *L. biguttata*, the only likely candidate for host status would appear to be *Esperia sulphurella* (F.) (Oecophoridae). However, this latter association is as yet unproven. A further rearing 'from *Artemesia* roots' most probably links to a host pupating in the soil (which could of course be *O. brumata*). An additional geometrid rearing is via *Alcis repandata*. The species is frequently taken in light traps.

Lissonota nitida Gravenhorst, 1829 [*agnata*]

Taxonomy: The species bears a passing resemblance to the common *L.* (*Meniscus*) *lineolaris* in terms of head structure, but is easily recognised from the characters given in the key.

Distribution, abundance and phenology: There are no UK records since 1942.

'Historical' data: 'Brit. Is.' (material from Capron / Morley / Stephens /Harwood). The most recent capture was: Suffolk, Linstead, 26.vi.1942. [NHM]. Ireland (O'Connor *et al.*, 2007): lists '*nitida* Bridgman', for which the correct name is *subaciculata* Bridgman (see p. 81). Flight period: June (so far as is known).

Biology: There are rearing records from *Synanthedon tipuliformis* (Cl.), thus the species should be sought in gardens and allotments (assuming of course, that the stated host association is valid).

Lissonota plana sp. nov. [*impressor*]

Female: Flagellum 1 from 6-7 x longer than broad; flagellum 2 greater than 4, to nearly 5 x longer than broad. Dorsal aspect of temples: contour strongly narrowing. Frons impressed medially, raised laterally. Interocellar distance about 1.3 x distance to eye, and about 0.8 x posterior ocellus to occipital carina – the latter rounded centrally. Sculpture: vertex with small punctures, narrower than interspaces; frons: as vertex, but punctures more distinct; face: coriaceous, with punctures subequal to interspaces. Clypeus 1.5 to nearly 2 x broader than long, with large punctures on basal band. Maximum temple length less than 0.66 x length of flagellum 1. Minimum length of gena about 1.3 x malar space – latter about 0.8 x basal width of mandible. Fore wing length: 9-10 mm. Mesopleurum with minute, deep punctures, these a little narrower than interspaces. Speculum 'open', large, extending about half way to epicnemium. Mesosternal sulcus trans-striate throughout. Hind tarsal claws with pectin to about 0.66. Propodeum strongly punctured, in part running to rugosity. Metasoma: tergite 1: nearly 3 x longer than broad; coriaceous, glabrous at apex, some puncturation before latter; length/breadth ratio about 2.5. Tergite 2 coriaceous, about 1.35-1.5 x longer than broad. Ovipositor around body length, or somewhat longer. Colour: body black; scutellum rarely with yellow lateral marks, mesonotum with or without indication of reddish shoulder marks; legs reddish, hind tarsus somewhat infuscate.

Male: similar to female, but flagellum 1 only 4-4.5 x longer than broad.

HOLOTYPE. SCOTLAND: female, Perthshire, Rannoch, ex *scoliiformis,* 3.vii.1911, (Cockayne) [NHM: 1919-288].

PARATYPES. two males, one female: 'British' (Harwood) [NHM]. **SCOTLAND:** three males, one female, Perthshire, Rannoch, 1911; female, same locality, 1928 (Cockayne). **IRELAND:** female, Wicklow, Woodenbridge, 21.vi.1950 (Faris) [NMS]; male, 'Irish, bred – ?host' [sic], 1905. (Cockayne).

Taxonomy: The name '*impressor*' was used by earlier authors for a complex of three species centred about *rufipes* of Brischke. The 'true' impressor now stands as a prior name for *basalis* Brischke 1865 (f. Aubert, 1978) – which in fact belongs to different species group. Aubert (1978) also lists *rufipes* Brischke as a synonym of *impressifrons* Thomson 1889. However the Thomson species cannot be distinguished amongst the *rufipes* complex on the basis of characters studied by Aubert. From my own study of type material, original descriptions, plus a good number of 'continental' specimens of '*rufipes sensu lato*', *L. plana* is not conspecific with either *rufipes* or *impressifrons*.

Distribution, abundance and phenology: *Lissonota plana* is rare in the UK, and no doubt of limited distribution owing to the identity of the host species. Flight period: during June (limited information – the existing data are both sparse and incomplete).

Biology: The species has been reared from *Synanthedon scoliaeformis* (Borkh.) (Scotland, Cockayne, NHM). The host is found on *Betula*, and is largely confined to woodland in parts of Wales, Ireland, and the Scottish Highlands.

Lissonota setosa (Geoffroy, 1785) Plate 9

Taxonomy: *L. setosa* is the largest British banchine, and one of the most impressive species in the family Ichneumonidae – the only other with which it might be confused being *L. fulvipes*. This insect has often been illustrated in popular books on Entomology, and casual photographs of large, broad ichneumonids with projecting ovipositor (usually non-banchines) tend to be labelled with its name.

Distribution, abundance and phenology: A reasonable number of specimens exist in older collections, but the species seems to have almost disappeared over the last half century or more – no doubt owing to the considerable contraction of the host's distribution and abundance.

'Historical' data: four specimens, 'Brit. Is.' (Desvignes); two, 'Britain' (Clifton); England: no data (Harwood); Huntingdonshire, Wood Walton, 1923. [NHM]; female: 'Cameron, 1907-114'; male, 'Morley, 8.vii.1902 Thornhill'; female, 'British Isles'. [NMS]. The only recent capture is of one female: Cambridgeshire, Ely, on a *Salix* bole containing *Cossus*, 2.viii.2012 [JPB].

Biology: The host is *Cossus cossus* (L.), living in the boles of various deciduous trees. Two related Nearctic species of the *Lampronota* group also attack Cossidae (Townes and Townes, 1978).

Given the extent to which the host's abundance and distribution have diminished over the last half century or more (plus the usual clandestine habits of *Lampronota*), it is hardly surprising that *L. setosa* is very rarely encountered in the field.

Lissonota fulvipes (Desvignes, 1856) Plate 10

Taxonomy: Morley synonymised the present species with the preceding, and the two are certainly closely similar (apart from the obvious difference in size). Both sexes can be discriminated following the characters given in the key.

Distribution, abundance and phenology: There are very few specimens in UK collections.

England: type plus five paralectotypes (Desvignes); Suffolk, 1938, 'walking over a willow trunk'; 'from *crabroniformis* in willow' (Hawkins), 1938 [NHM]. There are only two recent captures: female: Middlesex, Harefield Wood, amongst *Salix*, 28.v.1980 [JPB]; Yorkshire, Derwent Ings; 29.vi.87 [WE.]. Flight period: late May, through to the end of June (from scant existing data).

Biology: The host is *Sesia*. The most likely target, *S. bembeciformis,* is a common and widely distributed species.

Lissonota dormitor sp. nov.

Female: around 35 flagellar segments. Flagellum 1 length/breadth ratio: 5 to 6 x. Temples in dorsal aspect strongly narrowing and barely convex behind eyes; occipital carina weakly angled. Interocellar distance nearly 2 x distance between ocellus and eye, approximately equal to or a little greater than distance to occipital carina. Maximum temple length about 0.7 x flagellum 1. Minimum genal length greater than 1.5 x malar space, and about 0.66 x flagellum 2. Malar space 0.75, to approximately equal to basal width of mandible. Fore wing length 6-7 mm. Mesopleural sculpture: punctures mostly about equal to, in part narrower or greater than interspaces. Speculum glabrous, extending less than 0.33 x distance to epicnemium. Mesosternal sulcus with costae moderately developed, becoming stronger towards apex. Tarsal claws pectin with strong teeth, to about 0.66. Propodeum: dorsal longitudinal carinae evident only in region of area basalis, approximately parallel. Metasoma: tergite 1: length/breadth ratio: 1.25-1.35, heavily, closely punctate, including posterior band across more than 0.4-0.5 of dorsum; more or less striate over pre-apical impression. Tergite 2 with punctures at least equal to interspaces, and not less than 1.3 x longer than broad. Ovipositor longer than metasoma. Colour: black: head with small yellow dots on vertex; thorax with yellow shoulder marks extending to wing base, subalar prominence yellow – legs reddish testaceous, with front and middle coxae yellow-marked. Metasoma with dark ventral plica.

Male: Distinguished from female by having tergite 1 entirely punctate dorsally; flagellum 1 from 4-4.5 x longer than broad; interocellar distance 0.66, to approximately equal to distance from posterior ocellus to occipital carina. Maximum temple length about 0.8 x flagellum 1; malar space narrower than 0.66 x basal width of mandible.

HOLOTYPE ENGLAND: female, Ely (W.J. Cross), 24.vii.1890 [NCM].

PARATYPES. ENGLAND: two females, Berkshire, Windsor Forest, Malaise trap, 11.viii.-26.ix.1997; 3 males: Norfolk, Santon Downham, Malaise trap: heath with birch and pine, 17-29.vi.1983, 6-20.vii.1985 [NMS]. One male, Berkshire, Windsor Forest, Highstanding Hill, Malaise trap, old oak with *Juncus*, 14.vii.-5.vii.1995 [HM]; female, Berkshire, Windsor Forest, 9.viii.1930 (Donisthorpe); **WALES**: male, World's End, Clywd, ex fallen cones of *Abies procra*, coll. 6.ix.86 [NMS]. **SCOTLAND**: female, Midlothian, Dalhousie, ex *Dioryctria abietella* (D. & S.), Spruce, ix.1969; male – same data, vii.1969; male, E. Perthshire, Kinnaird, 11-18.ix.2007, Rothamstead M.V. trap [NHM].

Taxonomy: This species was originally discovered by Bridgman, although his manuscript key was never published. The name *'dormitor'* ("*dorma*" = sleep) reflects this history. Superficially, *L. dormitor* resembles the *Campocineta* section of *Lissonota*. The usual *Lampronota* subgeneric traits are nonetheless in evidence.

Distribution, abundance and phenology: The species has been found in conifer plantations in England and Scotland. It is known only from the type specimens listed above. Flight period: July, through to September.

Biology: *L. dormitor* has been reared from *Dioryctria abietella* [NHM], also from *Abies* cones [NMS] (full data given above). Host preference is obviously discordant with Lampronota. However, there is a rearing of a similar species from Sesiidae in *Alnus,* from the continent of Europe [NMS].

Lissonota pimplator (Zetterstedt, 1838)

Taxonomy: The specimens examined are similar to *pimplator* material from the European continent, although the latter are noticeably larger (fore wing lengths 7-8 mm, as against around 6 mm). From the known distribution of the host species recorded here, it is obvious that a different species must be the usual target for L. *pimplator* (see below).

Distribution, abundance and phenology: In the UK, known only from two reared specimens: one female: Scotland: Banff (Aberdeenshire), Tarlair, MacDuff, from cocoon of *Bembecia muscaeformis* (Esper) in *Armeria*, coll. 27.iii.2012, em. 8.v.2012; Cullen, ex *B. muscaeformis* in *Armeria maritima*, coll. in rootstock .x.2001, em. 29.v.2002 [NMS]. Flight period: no wild-caught material has been encountered.

Biology: The above mentioned sesiid host is associated with coastal regions, whereas on the European continent, *L. pimplator* has been reared through *Synanthedon spheciformis*. Disparities both in overall size plus host choice would appear to indicate at least subspecific status for the UK specimens. Analysis of variation in structural traits for *pimplator* from twenty specimens taken on the continent of Europe shows that all potentially specific characters exhibit a sufficient range of variation such as to counteract any move to propose full species status for the *Bembecia*-reared material. Two other *Lampronota* species are recorded in the present publication as having been recorded from more than a single host species (see L. *frontalis* and *canaliculata*).

Lissonota deversor Gravenhorst, 1829

Taxonomy: This species is very distinctive on the basis of colour pattern alone.

Distribution, abundance and phenology: Two historical records: one male: England: Devon, Staverton, 22.viii.18 (Morley coll.); one female: 'British' (Walker); one recent record: Herts., Aldbury, garden actinic light, 2.viii.2013 (G. Broad) [NHM]. Other material encountered by the present author had been misdetermined. Ireland (O'Connor *et al.*, 2007): cites data from Stelfox (this record requires confirmation, given the taxonomic information now available).

Biology: Unknown.

Lissonota freyi (Hellén, 1915), Plate 11

Taxonomy: *L. freyi* is easily recognised by the structure of the mesosternum. Males have tergite 1 more extensively striate than in females.

Distribution, abundance and phenology: Relatively little known in the UK – although probably more widespread than is indicated by the small number of existing records. England: Berkshire, Streatley, 'ex *andrenaeformis*', 1959 [NHM]; Hertfordshire, Doward, ex borings of *Synanthedon andrenaeformis* (Lasp.) in *Viburnum lantana*, coll. iv.1986 [NMS]. Kent, Folkestone, Malaise trap [HM]; Surrey, Colley Hill, Reigate, ex *Viburnum lantana* stems collected 25.iv.2000. Flight period: June-July.

According to Townes and Townes (1978), *L. freyi* is of Holarctic distribution.

Biology: The host species cited above is linked to *Viburnum* in limestone grassland (see also remarks on *L. canaliculata*).

Lissonota frontalis (Desvignes, 1856)

Taxonomy: This species and the next are closely related, and care needs to be taken in view of infraspecific variation.

Distribution, abundance and phenology: As with the preceding species, *L. frontalis* is probably overlooked, and needs to be sought by rearing potential host material. Existing records extend to Scotland. Few examples have been taken since the 'historical' period.

Four paralectotypes: England: 'Cornwall' / 'Brit. Is.' (Desvignes); further material: no data (Harwood); four – one labelled 'Colchester' (Harwood); one labelled 'Brit. Is.'; another: 'Brit. Is.' (Desvignes). Scotland: Aberdeenshire, Braemar district, 'ex *culiciformis*' [NHM]. One female: Dorset, Bourne Bottom, Parkstone, ex *Synanthedon formicaeformis* (Esp.), 24.v.1987 [NMS]. Flight period: no wild-caught data, emergence of reared material occurs during late May.

Biology: *L. frontalis* has been reared from *Synanthedon formicaeformis* (Esp.) in *Salix viminalis* – and apparently also from *S. culiciformis* (L.).

Lissonota canaliculata (Szépligeti, 1899) [pimplator]

Taxonomy: The present species was incorrectly identified as *L. pimplator* of Zetterstedt by Morley and other authors.

Distribution, abundance and phenology: Examples have been reared from Sesiidae in *Viburnum* and *Salix*, and the species may well occur wherever the hosts are to be found.

England: Kent, Folkestone, 1907. Six males, some labelled 'Redhill' (Lyle); three females, 'Bankes' (Morley); two males, two females – 'Kent / *andrenaeformis*' (Morley); Kent, Westerham, ex *Synanthedon andrenaeformis* (Lasp.); ex *S. formicaeformis* (Harwood); many females (Harwood). [NHM]. Seven males, three females, Kent, Folkestone, Malaise trap [HM]. Also: Surrey, White Downs, Gomshall, ex *V. lantana* stems collected 30.iv.2000. Flight period: August, possibly extending into early September.

Biology: The hosts, *Synanthedon andrenaeformis* (Lasp.) and *formicaeformis* (Esp.), are linked respectively to *Viburnum* and *Salix*. Habitats include damp locations with *Salix viminalis*, in addition to the usual downland *Viburnum* biotope (see also *L. freyi*).

Footnote: Kuslitsky (1981) uses the name *canaliculata* Szpl. for *frontalis* as defined in the present work. The fact that the type specimen of *canaliculata* is to some extent intermediate between these two segregates (plus the existence of some data implying an at least partially overlapping host range) indicates that further research is needed in order to ascertain whether or not we are dealing here with wide infraspecific variation within a single species. Similarly, *canaliculata* and *freyi* have been collected together at a single time and location. Despite a seemingly abrupt division between the latter two species in both sexes, additional (and ideally *reared*) material will be needed in order to fully assess the situation.

Lissonota subgenus *Lissonota* s. str

This subgenus contains some of the most abundant larger *Lissonota* species. Hosts are Noctuidae, including common grassland species. The group is regarded by (Townes and Townes, 1978) as being more closely related to *Alloplasta* and *Arenetra* than to other *Lissonota*. As with the UK, there are many very common Nearctic species – but very few rearings.

Key to species for subgenus *Lissonota* s. str

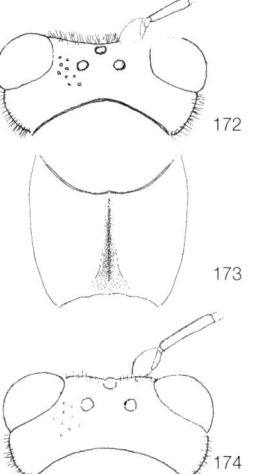

1. Length of hair on head (especially on frons) about 1.5 x ocellar diameter (Fig. 172); fore wing length about 10 mm; maximum temple length approximately equal to length of flagellum 1 – latter about 3.5 x longer than broad; mesosternal sulcus deeply excavate (Fig. 173); (metasoma entirely black; malar space approximately equal to basal width of mandible) *On* Luperina nickerlii, *sand dunes* ....
.... *sabulosa* **sp. nov.** (p. 70)

– Hair on head shorter than ocellar diameter (Fig. 174); (wing length generally less than 10 mm; maximum temple length much less than flagellum 1 – except when metasoma broadly red-marked; flagellum 1 at least 4 x longer than broad; (mesosternal sulcus often shallow; malar space often narrower than basal width of mandible) 2

2. Hind tarsal claw pectinate, the teeth usually on basal 0.5 or more; pleural carina usually crossing vertical groove – stronger than adjacent transverse costae (Fig. 175); (speculum larger on average; metasoma usually entirely black) 3

175

– Hind tarsal claw pectinate on only the basal 0.2 or less (and often in the form of bristles); pleural carina not crossing vertical groove, or else merging with adjacent costae (at least in females) (Fig. 176); (speculum smaller on average; metasoma usually at least partly red-marked) 4

176

3. Mesosulcus distinctly excavate (cf. Fig. 177). Male with fore and middle coxae and trochanters usually yellow-marked (often profusely so). *Abundant* .... *fundator* (Thunberg) [*sulphurifera*] (p. 70)

– Mesosulcus in the customary form of a shallow trough, somewhat deepening towards apex. Male with fore and middle coxae and trochanters unicolorous. *Widespread, sometimes common on grassland* .... *impressor* Grav. [*basalis*] (p. 70)

177

4. Median groove of mesosternum normal (i.e. slightly deepening towards apex); propodeum more sparsely punctate laterally than dorsally; minimum genal length at most 0.66 x length of flagellum 1; tergite 1 transversely coriaceous, with scattered puncturation in more lateral region; tergite 2 more or less coriaceous with scattered punctures towards basal angles (Fig. 178). Metasoma often dark suffused on central tergites. *Abundant grassland species* .... *clypeator* (Grav.) [*cylindrator*] (p. 71)

178

– Median groove of mesosternum deeply excavate (Fig. 179); propodeum uniformly punctate; minimum genal length often greater than 0.66 x flagellum 1; tergite 1 strongly punctate for the most part; tergite 2 closely and deeply punctate (Fig. 180). Metasoma usually broadly red (at least centrally). *Uncommon species* 5

179

5. Metasoma apically black in females, larger (fore wing length 7-8 mm). *Widespread, not restricted to high altitudes* .... *digestor* (Thunberg) (p. 71)

– Metasoma entirely red in females, smaller (fore wing length 6-7 mm). *Uplands, including high montane* .... *magdalenae* Pfankuch (p. 72)

180

Species accounts for subgenus *Lissonota* s. str.

Lissonota fundator (Thunberg, 1824) [*sulphurifera*]

Taxonomy: This is the commonest *Lissonota* species having a deeply excavate mesosulcus. The male sex may be confused with *L. impressor*, but can usually be separated on leg colour pattern. The species is referred to under the name '*sulphurifera*' in the earlier literature.

Distribution, abundance and phenology: Very widely distributed and common. Ireland (O'Connor *et al.*, 2007): widespread. Flight period: adults appear during July-August.

Biology: Frequents grassy localities such as commons and open ground in general, where it attacks larvae of the *Mesapamea secalis* aggr. (L.) [NHM]. The species has also been reared from *Luperina testacea* [NCM]. Adults are commonly found on flowers of umbellifers, and sometimes appear in light traps.

Lissonota impressor Gravenhorst, 1829 [*basalis*]

Taxonomy: This is the '*L. basalis*' of the older literature. There is occasional near overlap between extremes of this species and the previous with regard to form of the mesosulcus. For this reason, it is always best to collect many specimens together, in order to facilitate comparative assessment.

Distribution, abundance and phenology: Widely distributed – although less common than *fundator*. Habitats are similar to those of the latter species. Ireland (O'Connor *et al.*, 2007): widespread (in this publication, reference to an apparently undescribed species refers to an earlier misidentification of *impressor* Grav., and not to the present species (see *canaliculata* Szépligeti, p. 68)). Flight period: July and August.

Biology: *L. impressor* is found in grassy localities, the adults often occuring on umbellifer blossom. Rearings: two females, one male, Looe Bar, Cornwall, 'ex roots of *Elymus farcus*, with *L. nickerlii*'; also: one male in feeding tube, *L. nickerlii*'. [NMS].

Lissonota sabulosa sp. nov.

Female: Flagellum 1 about 3.5 x longer than broad. Head pubescence about 1.5 x longer than ocellar diameter. Temples in dorsal aspect with rounded contour. Interocellar distance subequal to distance between posterior ocellus and compound eye, and 0.8 x posterior ocellus to occipital carina. Vertex coriaceous and strongly punctured (except at hind ocelli). Frons and face as vertex. Maximum temple length approximately equal to length of flagellum 1. Malar space approximately equal to basal width of mandible. Minimum genal length about 1.3 x malar space. Fore wing length about 10 mm. Mesopleurum with strong puncturation, punctures mostly larger than interspaces. Speculum very small – mostly obscured by punctures (the latter smaller than interspaces). Mesosternal sulcus deeply excavate behind, transversely striate throughout. Hind tarsal claw with pectin of bristles to about 0.4 x length of claw. Propodeum with large, deep punctures, these mostly larger than interspaces. Metasoma: tergite 1 nearly 2 x longer than broad, closely and deeply punctured, with some longitudinal rugosity laterally. Tergite 2 subquadrate. Ovipositor: greater than body length. Colour: black-bodied, with legs red – the coxae and trochanters and trochantelli black.

Male: unknown.

HOLOTYPE. ENGLAND: female, Kent, Sheppey – ex *Luperina nickerlii* – coll. 1990, parasitoid adult emerged 4.iv.1991.

PARATYPE. ENGLAND: female: same data as previous – emerged 10.iv.1991 [NMS].

Taxonomy: This is the largest of the *Lissonota s. str.* species, at once recognisable from the hirsute appearance of the unusually buccate head, plus an apparent link with a sand dune habitat.

Distribution, abundance and phenology: Known only from the type series.

Biology: The host is *Luperina nickerlii* (Freyer), feeding on roots of duneland Poaceae. Whether this implies a specific relationship with this noctuid species is not known. It must be borne in mind that the related *L. impressor* has been reared, both from this host and from other noctuid species living in different biotopes.

Lissonota clypeator (Gravenhorst, 1820) [*cylindrator*] Plate 12

Taxonomy: The reddish central metasomal tergites generally serve to separate typical specimens of this species from other common *Lissonota* taken in similar biotopes.

In earlier literature, the species is generally referred to under the name *cylindrator*. Townes and Townes (1978) treat eight subspecies for the Nearctic region.

Distribution, abundance and phenology: Widely distributed and common, more or less throughout the UK. Ireland (O'Connor *et al.*, 2007): records from counties Down and Louth. Flight period: August to September.

Biology: *L. clypeator* frequents grassy localities of the same kind as the last two species, the adults frequently feeding on umbellifer flower tables. Hosts: *Oligia* sp. (Norfolk, Giston [NMS]); one example 'ex *A. secalis*', (L.) [NHM]; another from the latter host: *secalis* larva collected mid-June, parasitoid larva emerged last week in June. The adult ichneumonid then formed within about two weeks, but remained dormant until October, when it finally emerged [JPB].

American subspecies of *L. clypeator* have been reared from *Septis* (= *Apamea*) *devastator*, and from an unidentified 'cutworm' (Townes and Townes, 1978).

Lissonota digestor (Thunberg, 1822)

Taxonomy: The unusually heavy puncturation of the metasomal tergites will readily separate this and the next species from the general milieu of *Lissonota* sensu stricto. *L. digestor* sometimes has a predominantly infuscate metasoma.

Distribution, abundance and phenology: Rather scarce in collections, although very widely distributed in the UK – extending northwards at least to Inverness-shire.

England: Oxfordshire, Barrow Farm Fen, Malaise trap, 5.iv.- 17.v. [NHM]; Wiltshire, Braydon, Red Lodge Pond NR. Scotland: Creag Meagaidh, 375m, Malaise trap, 15.v.-17.6.; 3-16.v.; Dunbartonshire, Gartlea, Caldarvan, 31.v.; Perthshire, Crianlarich; Angus, Glas Maol [NMS]. These data indicate that adults have an unusually early flight period.

Biology: The host is *Gortyna flavago* (D. & S.), feeding internally in thistle stems (Bridgman Collection [NCM] – also following Aubert (1978), f. Habermehl). Habitats may extend to montane regions, although *L. digestor* is perhaps most often encountered in lowland habitats.

Lissonota magdalenae Pfankuch, 1921

Taxonomy: Closely related to the preceding species, although of smaller average size. In addition, *magdalenae* seems to be a predominantly montane species, whereas *digestor* is much less restricted in distribution. Usually, both sexes of *digestor* can be distinguished from the present by having the apical tergites blackish – but this is not always reliable in males. From the data given below (and under the previous species), it is seen that *digestor* and *magdalanae* have been collected at the same time, in the same location. Given that separation is based mainly on size and colour, it may well be that we are looking at infra-specific variation within a single species. There are intermediates between the two, thus the possibility remains that *magdalenae* is a montane form of *digestor* (possibly attacking a different host).

Distribution, abundance and phenology: Probably not rare at high altitude on Scottish mountains, distribution extending as far north as Sutherland and Wester Ross.

England: Yorkshire, Pen-y-Ghent, 'about 2000 ft'. Scotland: Sutherland, Inchnadamph NR, Ben Eighe [NHM]. Wester Ross, Anteallach, Blanket Bay, 270m, MW bowl trap, 7.v.–17.vi.; Cairngorms, Ben MacDui, on snow, 1.vi.; Westerness, Ben Nevis summit [NMS]. Perthshire, Glas Maol, 3000 ft, 25.vi. (UM). Ireland (O'Connor *et al.*, 2007): Donegal; Wicklow (1700 ft altitude).

Biology: Host unknown. The biotope seems usually to be high altitude open ground, although I have personally encountered the species on conifer foliage at lower levels (Perthshire, Aberfoyle).

Lissonota subgenus *Loxonota*

This subgenus resembles *Lissonota* s. str. in that the tarsal claws tend to be long in relation to the arolium. However, the sternal plicae are pale rather than dark in colour, and the fore wing venation is distinctive. There are no common or widely distributed species in this subgenus, so far as the UK is concerned. *Loxonota* species appear to inhabit sandy localities, including both coastal dunes and breckland. *L. cruentator* and *histrio* males have distinctive yellow longitudinal markings on the mesonotum (absent in *lineata*). The thorax may also be partly suffused with reddish colouration – and males may have the coxae richly yellow-marked. Fore wing lengths 5-7 mm.

The recent revision by Rey de Castillo (1992) lists 8 species for the Western Palaearctic region.

Key to species for subgenus *Loxonota*

1. Female 2

– Male 4

2. Ovipositor conspicuously downcurved apically (Fig. 181); central tergites more elongate (tergite 2 longer than broad); tergite 1 with punctures extending across dorsum (sometimes glabrous in small part) (Fig. 182); occipital carina distinctly angled centrally; areolet met by second recurrent vein before middle; flagellum with at least 45 segments .... *cruentator* (Panzer) [*insignita*] (below)

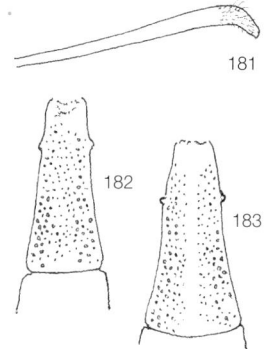

- Ovipositor approximately straight throughout; central tergites less elongate (tergite 2 broader than long or only slightly longer than broad); tergite 1 with glabrous median line (Fig. 183); occipital carina rounded, dipped or squared centrally; areolet met by second recurrent vein beyond middle; flagellum with less than 45 segments 3

3. Tergite 3 subquadrate to broader than long, more or less dark-suffused; female with flagellum 1 at most 4 x longer than broad; propodeum with a dent in position of area basalis; mesonotum entirely piceous .... *lineata* Grav. (p. 74)

- Tergite 3 (usually also 2) longer than broad, reddish (sometimes slightly dark suffused); female with flagellum 1 at least 4 x longer than broad; propodeum lacking basal dent; mesonotum with yellow longitudinal markings .... *histrio* (F.) [*parallela*] (p. 74)

4. Minimum genal length about 0.33 x eye length; occipital carina with at most a weak central angulation; frons trans-striate medially; tergite 1 impunctate dorso-medially; propodeum extensively, transversely striate .... *lineata* Grav. (p. 74)

- Minimum genal length less than 0.33 x eye length; occipital carina sharp-angled centrally; frons, tergite 1 and propodeum uniformly punctate 5

5. Inter-ocellar space about 2 x distance between posterior ocellus and eye, and clearly greater than distance between posterior ocellus and occipital carina; coxae richly yellow-marked .... *histrio* (F.) [*parallela*] (p. 74)

- Inter-ocellar space about 1.3 x posterior ocellus to eye, and subequal to distance between posterior ocellus and occipital carina; coxae at most with restricted pale markings. .... *cruentator* (Panzer) (below)

Species accounts for subgenus *Loxonota*

Lissonota cruentator (Panzer, 1809) [*insignita*]

Taxonomy: *L. cruentator* is unlikely to be mistaken for any other British *Lissonota*, on account of the highly unusual ovipositor structure, combined with presence of yellow lines on the mesonotum in males. This species is the *L. insignita* of older literature.

Distribution, abundance and phenology: A rarely encountered species, *L. cruentator* is probably a predominantly coastal insect – although a few inland records do also exist. Distribution is apparently confined to the southernmost counties.

England: one male, one female, Kent, Graveney Marshes, on *Daucus carota* flowers, 17.viii.2009; Surrey, Box Hill; nr. Mitcham (Desvignes); Isle of Wight, St. Helens; Jersey, St. Catherines, several on *Daucus* [NHM]. Female: Hampshire, Hayling Island, M.V. light, 4.vii.; female, S. Devon, Dawlish Warren NNR, 31.vii.; male, ditto -, ex *Synaphe punctalis* (F.) on *Hypnum sp.*, coll. 10.iv. (Morley gives the same host, teste Bridgman) [NMS]. Ireland (O'Connor *et al.*, 2007): Mayo (teste Morley).

Biology: The known host is linked to coastal duneland habitats.

Lissonota lineata Gravenhorst, 1829

Taxonomy: There is some overlap between the present species and *L. histrio*, although the two should be separable on characters given in the key.

Distribution, abundance and phenology: There are but few British records for *L. lineata*. As with the previous species, coastal or other sandy locations seem to be preferred.

England: Kent, Deal, (Desvignes); E. Sussex, Camber Sands, on *Daucus carota*. [NHM]. Six males, nine females; Norfolk, Santon Downham, Malaise trap, heath with *Betula* and *Pinus*, 1-15.vii.; Lancashire, Ainsdale; Norfolk, Horsey. Wales: Clwyd, Gronant [NMS]. Flight period: taken through July and August.

Biology: The only confirmed host is *Anerastia lotella* (Huebn.) (from material collected at Grovant, NMS). This lepidopteran is of very local distribution in Britain, preferring coastal locations, but occurring also in breckland.

Lissonota histrio (Fabricius, 1798) [*parallela*], Plate 13

Taxonomy: This is the *L. parallela* of the older literature.

Distribution, abundance and phenology: As with the other two British species of the *Loxonota* group, *L. histrio* is of rare occurrence in the UK. However, while *cruentator* and *lineata* appear to be of strictly southern distribution, *histrio* records extend to Scotland.

England: a number with no data (Desvignes, Marshall, Morley). Several (including a female found under a stone), Suffolk, Lowestoft (Morley); ten specimens, Essex, Clacton, (Harwood) [NHM]; one female, Norfolk, Horsey; one male, one female, 'Holme' (Morley); one female, Norfolk, Santon Downham; Scotland: Fife, Tentsmuir [NMS]. Ireland: Curracloe, (Stelfox) [NHM], one female, Ballyteige; O'Connor *et al.* (2007): Dublin; Mayo; Wexford; Wicklow. Flight period: July and August.

Biology: Aubert (1978) lists several reputed hosts, whereas Rey de Castillo (1992) gives none. The present is apparently a coastal species (although with one record from Breckland).

Lissonota subgenus *Campocineta*

The majority of *Lissonota* species belong in this section. The species groups of *Campocineta* are to a fair extent artificial. Where exceptions to a given segregate exist, these are cross-referred between different sections of the subgenus. Townes and Townes (1978) regard the *coracina* group as being related to the genus *Loxodocus* – which latter Aubert (1978) considers probably synonymous with *Lissonota*. The remainder of Nearctic *Lissonota* are placed in species groups that are more or less arbitrary, with any attempt to apply these partitions to non-Nearctic species proving unsatisfactory or confusing (as indeed predicted by Townes). With regard to taxonomic data, it should be noted that the Townes characters involving distribution of setae on the metasomal laterotergites and tergites are subject to much modification by abrasion. Consequently, I have generally used these traits sparingly. Lastly, it should be said that some males will fail to run to species group. I have attempted to render this problem as non-threatening as possible. However, the only safe approach is to rear or collect the sexes together.

Key to species groups of *Campocineta*

1. Claws nearly 2 x length of arolium; metanotum about 2 x distance between petiolar spiracles; sternites and ventral plica dark; (metasomal tergite 1 with transverse microsculpture; legs red, tibiae and tarsi dark). (*Common on dead wood*) .... **biguttata** Holmgren (p. 63)

\- Claws at most a little greater than 1.5 x length of arolium; width of metanotum usually less than 2 x distance between petiolar spiracles; sternites usually pale, ventral plica always so; (metasoma lacking transverse microsculpture and / or legs with different colour pattern) 2

2. Areolet narrowly pentagonal; ovipositor somewhat scimitar-shaped and about length of metasoma (Fig. 184), first and second metasomal tergites predominantly coriaceous, with indistinct sculpturing other than on postpetiole; mesonotum black with red markings, scutellum red laterally; minimum genal length about 1.5 x malar space; speculum small, but distinct; legs entirely red. *Montane species, Scotland* .... **erythrina** Holmgren (p. 78)

\- If rarely the areolet is pentagonal, then disagreeing in at least two other characters 3

3. Femora usually spindle-shaped (swollen towards base and noticeably constricted distally (Fig. 185)). Males with abnormally long hair on head and thorax (distinctly longer than maximum width of hind ocellus); minimum temple length at least 0.5 x eye length, gena strongly widening below (Fig. 186); speculum moderately large and shining. *Montane species, Scotland, Derbyshire* .... **admontensis** group (p. 78)

\- Femora of normal build; males lacking long hair on head and thorax; minimum temple length rarely approaching 0.5 x eye length, gena usually narrowing below 4

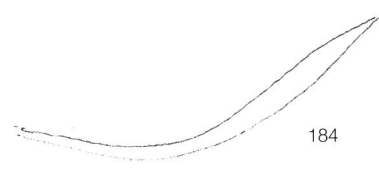

184

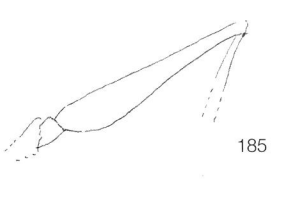

185

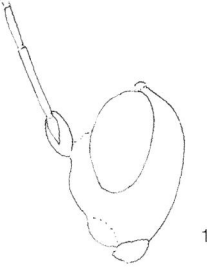

186

4. Flagellar segments broader than long in distal region in female sex; areolet (cell 2Rs) open externally, or with the outer vein much reduced; propodeum heavily rugose on dorsum (Fig. 187); mesonotum with notauli long and distinctly impressed (Fig. 188); face with distinct central keel (Fig. 189); gena somewhat widening towards venter (Fig. 190); (central tergites coriaceous and impunctate, at least partly reddish suffused in females, legs reddish, largely black in males) .... *linearis* group (p. 79)

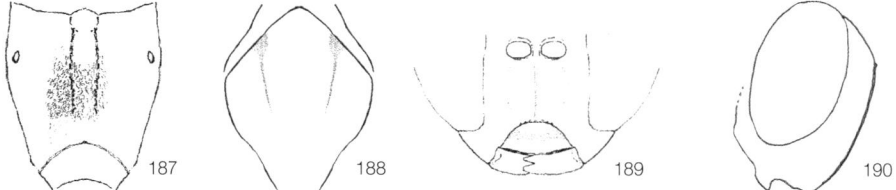

– Flagellar segments subquadrate to longer than wide in distal portion (only the last 2 or 3 sometimes broader than long); areolet usually complete (or nearly so); propodeum rarely heavily rugose dorsally; mesonotum with at most short, shallow notauli; face rarely with central keel 5

5. Metasoma with central tergites predominantly red; laterotergite 3 with less than 6 hairs (often bare); area of speculum with large punctures, wider than those on adjacent mesopleurum (Fig. 191), (tibia 3 not yellow at base) .... *coracina* group (p. 80)

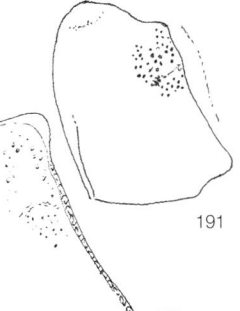

– Central tergites predominantly dark (if one or more tergites wholly or mainly red, laterotergite 3 extensively hairy; speculum often large (e.g. Fig. 192), impunctate and shining, tibia 3 sometimes yellow at base) 6

6. Clypeus with unusually dense hair (Fig. 193); (laterotergite 4 completely inflexed (Fig. 194); central tergites closely, deeply punctate (Fig. 195); fore wing length at most 4 mm) .... *clypealis* group (p. 82)

– Clypeus with normal vestiture – or if relatively dense, laterotergite 4 not completely inflexed and central tergites not, or less closely punctate 7

7. Postpetiole with striate trans-impression – *tergite 2 generally also with a similarly sculptured pre-apical zone* (absent in some males) (Fig. 196); ovipositor weakly to strongly expanding towards apex; (*fore wing length 4-5 mm*; maximum width of pterostigma often not less than 0.5 x radial sector (Fig. 197); flagellum usually with less than 30 segments) .... *saturator* group (p. 83)

– Tergite 1 sometimes with a striate trans-impression – if rarely, with similar sculpture on tergite 2, then ovipositor not expanding towards apex; (*fore wing length usually in range 5-7 mm*; pterostigma often less than half radial sector; flagellum often with more than 30 segments) 8

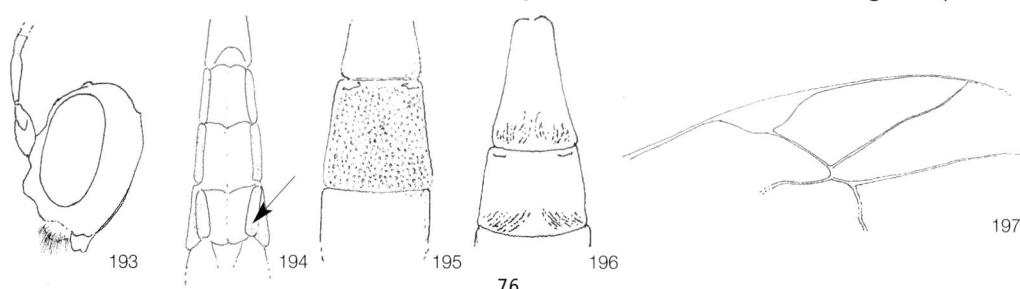

8. Tergites 2 and 3 of metasoma with breadth 1.4-2 x greater than length, entirely reddish pigmented .... cf. ***fletcheri*** [*saturator* group] (p. 87)

–   If the central tergites are strongly transverse, then with different colour pattern 9

9. Tergite 1 and central tergites uniformly coriaceous, former with pre-apical impression at most longitudinally wrinkled – latter with at most a few scattered punctures in basal half. Tergite 2 subquadrate to distinctly elongate (Fig. 198); laterotergite 3 bare to moderately hairy (Fig. 199) tergites 5 and 6 usually shining and with few dorsal hairs (Fig. 200); mesosulcus evenly (and often strongly) trans-costate throughout; speculum usually small and / or predominantly coriaceous .... ***gracilenta*** group (p. 90)

–   Tergite 1 at least partly punctate and / or striate (especially over pre-apical impression) and central tergites usually copiously punctate, including posterior part – except when tergite 2 is distinctly broader than long; laterotergite 3 with numerous hairs (Fig. 201); (also tergites 5 and 6 often with conspicuous microsculpture and usually with many dorsal hairs (Fig. 202); mesosulcus with transverse costae usually well developed only in posterior portion (or else largely obsolete; speculum usually large and at least partly shining) 10

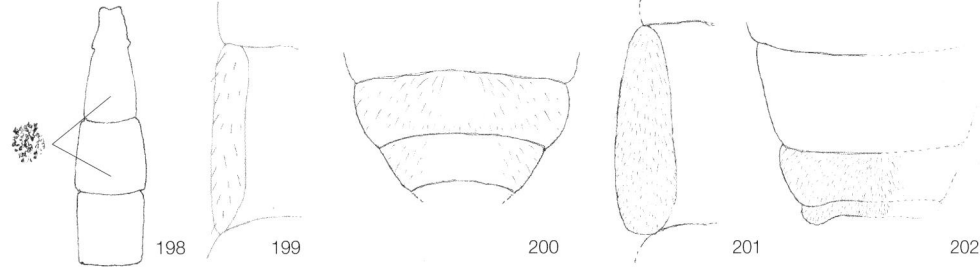

198   199   200   201   202

10. Mesonotum and / or mesopleurum predominantly red; (metasoma predominantly black; tergite 2 partly to entirely punctate, nervellus well developed) .... ***versicolor*** group (p. 92)

–   Thorax predominantly black (if red-marked, metasoma mostly red, central tergites coriaceous medially, and nervellus weakly developed) 11

11. Head large, 'sub-globose' in dorsal aspect – long axis of head, half to nearly equal to external distance between compound eyes in dorsal aspect (Fig. 203): propodeum dorsally coriaceous and dull, at most weakly punctured in frontal part of area superomedia; tergite 1 predominantly coriaceous (Fig. 204); *fore wing length less than 5 mm.* [Male with broad yellow apical margins to tergites 2-4, coxae 1 and 2 yellow, face, pronotum, shoulders yellow marked]. *Widespread. Associated with conifers* .... ***dubia*** group (p. 94)

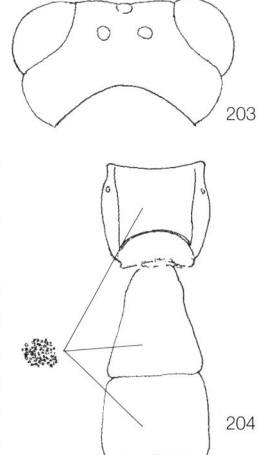

203

204

–   Head more transverse – long axis of head usually distinctly less than half external distance between compound eyes; propodeum generally extensively punctate; tergite 1 usually at least in part punctate and / or striate (exceptions generally having fore wing length greater than 5 mm). *Medium-sized species, mostly in deciduous woodland* .... ***buccator*** group (p. 94)

Species accounts for subgenus *Campocineta*

Lissonota erythrina Holmgren, 1860

Taxonomy: *L. erythrina* is unlikely to be mistaken for any other British *Lissonota*, the combination of colour pattern plus ovipositor structure being quite distinctive.

Distribution, abundance and phenology: The species is known in the UK from a single example taken at high altitude in Inverness-shire.

One female: Scotland: Inverness-shire, Creag Meagaidh, 760 m, *Vaccinium hybridella / Myrtillus* heath, Malaise trap, 18.vi.-10.vii. [NMS].

Biology: *Lissonota erythrina* is clearly a high montane species. The host is unknown, but should be sought amongst 'micro' Lepidoptera living in such environments.

Key to species for *admontensis* group

Both species in the *admontensis* group have the central tergites at least partly reddish; tergites 3-4 tend to have lateral black areas. The unusually long pubescence is very noticeable in males, but much less so in females (for which reason, Aubert (1978) at first confused the female of *admontensis* with *L. proxima* of the *coracina* group). The hind femora are slightly, to strongly spindle shaped in both species.

1. Gena nearly 2 x width of mandible base, mandible predominantly piceous; tergite 2 longer than broad; larger species – wing length up to 6 mm; hind femora noticeably spindle-shaped (Fig. 205); temples more narrowing in dorsal view (Fig. 206); ovipositor distinctly longer than metasoma. *Scotland* .... ***genator*** Aubert (below)

−   Gena at most 1.7 x width of mandible base, latter predominantly yellow; tergite 2 subquadrate to transverse; smaller species – wing length distinctly less than 6 mm; hind femora normal, to moderately spindle-shaped (Fig. 207); temples somewhat swollen in dorsal aspect (Fig. 208); ovipositor at least equal to metasoma plus propodeum. *Derbyshire, Scotland* .... ***admontensis*** Strobl (p. 79)

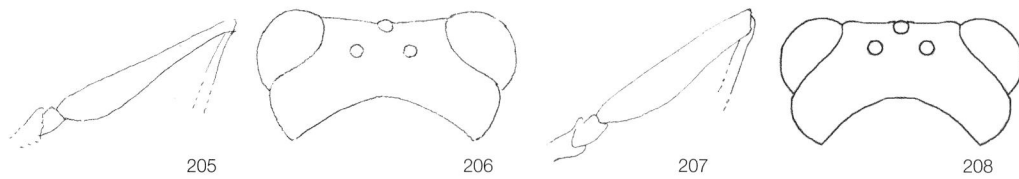

205   206   207   208

Species accounts for *admontensis* group

Lissonota genator Aubert, 1972

Taxonomy: Males of this and the next species are instantly recognisable on account of the long hair on the head, the 'buccate' form of which latter is diagnostic for both sexes (provided that other characters are also taken into account).

Distribution, abundance and phenology: Known in the UK from a relatively small number of specimens taken at high altitude in the Scottish Highlands.

Scotland: male, female, Inverness-shire, Craig Meagaidh NNR; Coire nan Laogh, 780 m, 7-23.v.1988; Cairngorms, Coire Brochan, 1050 m, 15.vi.->14.vii.; Affric Hills, 850 m, water trap, 1-22.vi.; male, female: Angus, Glas Maol, on snow, 29.v. 2002 (Bland); male: Argyllshire, Bein Odhar, Tyndrum, snowfield, 800 m, 17.iv.1982 [NMS].

Biology: *L. genator* is obviously a montane species, apparently associated with grassland-heath. There is a single rearing from larva (possibly pyralid) in silk tunnel, *Thymus*, Perthshire (N00180), host coll. vii.1999, parasitoid emerged iv.2000 (K. Bland) [NMS].

Lissonota admontensis Strobl, 1902

Taxonomy: *L. admontensis* resembles a smaller version of *genator*.

Distribution, abundance and phenology: Known in the UK from a few specimens taken in the Scottish highlands, and in the Peak District, Derbyshire.

England: Derbyshire, Glossop, *Vaccinium* moor, 29.iv. [NMS]. Scotland: Angus, Glas Maol, one female, snowbeds, 7.v.; one female, same location as previous, 20.iv.; one female, Angus, Glas Maol, White bowl, tall herbs at cliff edge, 28.vi.-11.vii.; Inverness-shire, Creag Meagaidh NN, *Deschampsia* grassland, Malaise trap, 16.v.-18.vi.; Ross and Cromarty, Fannich Hills, SSSI, Meall Gorm, on summit ridge, *Rhacomitrium* heath, 10.iv..

Biology: This is another high altitude species, clearly associated with a similar biotope to *L. genator* – and also having no known host relationships.

Species accounts for *linearis* group

Lissonota linearis Gravenhorst, 1829

Taxonomy: *Lissonota linearis* superficially resembles the common *L. proxima*, and might easily be overlooked on that basis. However, the antennal structure is quite distinctive, and other characters given in the key leave no room for confusion with the *coracina* group.

Distribution, abundance and phenology: Known from a very few older records, supplemented by some recent rearings via Psychidae.

One example labelled 'Whittle Coll.'; second example: 'Cracknore Hard, Fasnidge' (Hampshire) [NHM]. Recent records: England: one female, Essex, Chigwell Row, tree bark, ex *Epichnopteryx plumella* (D & S), coll. 3.iv.2003; one female, Essex, Fobbing Salt Marsh, ex *Whittleia retiella* (Newm.), 17.iv.2002; one male, ex *E. plumella*, 31.iii.2003; one male, Essex, Hainault Forest, Chigwell Row, ex *E. plumella*, 16.iv.2009; one male, Kent, Trottescliffe, ex *E. plumella*, 25.iv.1969 [NMS]. Ireland (O'Connor *et al.*, 2007): Down (this determination requiring confirmation). Flight period: the only wild-collected specimen examined indicates May.

Biology: The Whittle specimen gives **'*E. reticella*'** [sic] as host – which clearly needs interpretation in terms of *lapsus calami*, given the other data now available. The psychid hosts are of very local and restricted distribution in the UK, and inhabit grassland, moorland and parkland.

Key to species for *coracina* group

This segregate is probably a natural one. Despite the abundance with which some species occur, host relationships are very poorly known.

1. Tergite 1 usually with heavier puncturation, plus a variable degree of striation over the postpetiole (Fig. 209); punctures on tergites 2 and 3 at least in part greater than or equal to interspaces; coxae black, femora at least partly black or dark-suffused; (fore wing length up to 6 mm). *Widespread, fairly common grassland species* ....
.... ***proxima*** Fonscolombe [*commixta*] (p. 81)

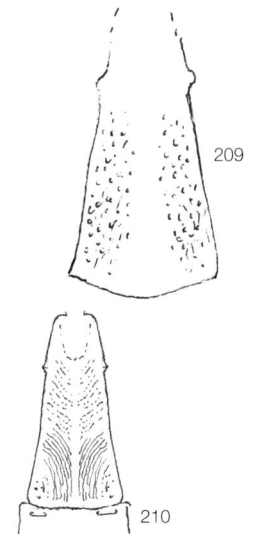

209

— Tergite 1 usually predominantly coriaceous, with some superficial punctures – only rarely with sharp, well-defined striae (Figs 210, 211, 212), tergites 2 and 3 generally more sparsely punctate – especially towards postero-median region; coxae and femora usually red, rarely black; (fore wing length rarely > 4 mm) 2

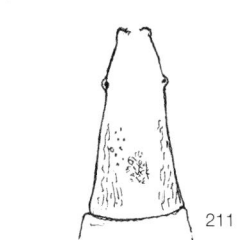

2. Tergite 1 with pre-apical striae extending diagonally over median part of petiole (Fig. 210); (distance between lower margin of compound eye and hypostomal carina at most 1.0 x length of flagellum 2). *Rare* .... ***subaciculata*** Bridgman [*nitida*] (p. 81)

210

— If tergite 1 with extensive pre-apical striae, then these are less steeply angled, and often more or less confined to post-petiolar region (lower margin of compound eye to hypostomal carina at least 1 x flagellum 2) 3

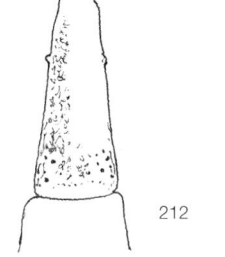

211

3. Postpetiole with extensive subapical striation which tends to extend onto the medial portion (Fig. 211). *Few UK records* .... ***argiola*** Grav. (p. 81)

— Postpetiole at most with restricted lateral striation (Fig. 212). *Abundant* .... ***coracina*** (Gmelin) [*bellator*] (p. 82)

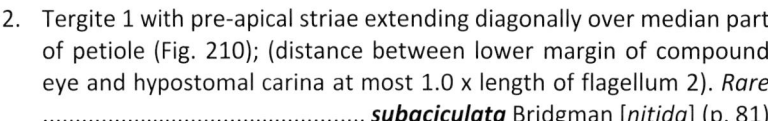

212

Remarks on identification of males:

The male of *subaciculata* is unknown, which makes it difficult to provide a satisfactory key. So far as the two common species are concerned, female structural characters will not separate *coracina* from *proxima*. However, the latter species has the coxae and trochanters piceous, whereas in *coracina* they are usually yellow – or else extensively yellow-marked (as is also, the mesopleurum). The usual size differences apply. With the rare *argiola*, the central metasomal tergites carry a broad apical yellowish band, and the hind coxae are piceous.

Species accounts for *coracina* group

Lissonota proxima Fonscolombe, 1854 [*commixta*]

Taxonomy: The head proportions of *proxima* are diagnostic in comparison to the other members of the species group. The sculpture of tergites 1-3 is subject to much variation and a certain degree of overlap with related species must be expected. The species was known under the name *commixta* in the older literature.

Distribution, abundance and phenology: Common and widely distributed. Ireland (O'Connor *et al.*, 2007): widespread. Flight period: May to September.

Biology: The biotope is meadow and grassland in general, including gardens. The host is unknown, but might possibly be sought amongst the *Crambus* group? 'Apocryphal' host records from Psychidae cited by Aubert (1978) no doubt involve confusion with *L. linearis*.

Lissonota subaciculata Bridgman, 1886 [*nitida*]

Taxonomy: The present species is unlikely to be mistaken for any other British *Lissonota*, on account of the distinctive sculpture of tergite 1.

Distribution, abundance and phenology: There are scattered records from southern England.

Holotype female, 'Gt Britain'. [NCM]. England: Hampshire. Suffolk, Monks Soham (Morley); Isle of Wight [NHM]; Hertfordshire, Doward; Dorset [NMS]. Ireland (O'Connor *et al.*, 2007 – see comments under *nitida* Grav. Earlier records require confirmation). Flight period: no wild-collected examples have been encountered.

Biology: *L. subaciculata* has been reared from *Mecyna asinalis* (Huebn.) [RSM, NHM]. The host species inhabitats open biotopes, such as meadow and downland.

Lissonota argiola Gravenhorst, 1829

Taxonomy: This species has an unusual history, in that it has been known only from the male sex for over a century. Identification of the female sex of *Lissonota argiola* is based partly on material from the European continent, loaned by Klaus Horstmann.

Distribution, abundance and phenology: Of somewhat rare and local occurrence in the UK.

England: three specimens from Desvignes, one from Stephens [NHM]. Yorkshire, Fen Bog NR; Salter's Brook [W. Ely, *pers. comm.*]; Kent, Folkestone Warren, Malaise trap [HM] Wales: (male) Angelsey, Llangristiolus, Malaise trap, 7-27.viii.1982; (female) Rhosgyll Fawr, Caernarfon, Welsh Peatland Survey, 14-28.vii.1988. Scotland: (female) East Dunbartonshire, Bardowie, vii.1872 (Cameron) [NMS]. Ireland: Dublin, Glenasmole, 13.vi.1933 [NHM].

Biology: *L. argiola* has been taken on open ground, including peaty heathland. Hosts are unknown.

Lissonota coracina (Gmelin, 1790) [*bellator*]

Taxonomy: This is an extremely variable species, particularly in the male sex. It should be noted that only a small proportion of females agree in antennal structure with Nearctic material referred to *coracina* by Townes and Townes (1978), so it is uncertain whether the latter is truly identical with the European species.

Distribution, abundance and phenology: Probably the commonest banchine in the UK, abundant everywhere; also Ireland (O'Connor *et al.*, 2007): widespread. Flight period: July to September.

Biology: *L. coracina* inhabitats open grassland habitats, including gardens and waste ground. The host is unknown (most probably one or more common *Crambus* sensu lato species). Morley gives *Mecyna flavalis* as a host (no doubt as a misdetermination of *subaciculata*). One Nearctic member of the *coracina* group, *L. imitatrix* Walsh, is stated to have been reared from '*Crambus* sp.' (Townes and Townes,1978).

Key to species for *clypealis* group

These are two small *Lissonota* species with closely punctate central metasomal tergites. The clypeal pile is distinctive, although 'comparative' in the sense that familiarity with the situation in other species of the genus will certainly aid identification.

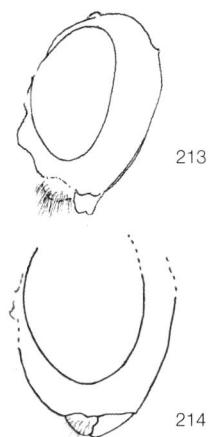

213

214

1. Minimum genal length greater than flagellar segment 2, and 0.35 to 0.4 x eye length (see Fig. 213); flagellum 1 about 4 -4.5 x longer than broad; flagellum 2 less than 2, to nearly 3 x longer than broad; clypeus medially flat and more densely pilose; frons, vertex, thorax and scutellum entirely black. *Widespread, uncommon* ....
    .... *clypealis* Thomson (below)

–   Minimum genal length much less than length of flagellar segment 2, and less than 0.35 x eye length (Fig. 214); flagellum 1 about 5 x longer than broad (4.5 in male); flagellum 2 about 3.5 x longer than broad; clypeus strongly convex and less densely pilose.; frontal orbits, vertex, thorax and scutellum with variable yellow markings. *Rare* ....
    .... *gracilipes* Thomson (p. 83)

Species accounts for *clypealis* group

Lissonota clypealis Thomson, 1877

Taxonomy: The clypeal vestiture should separate this and the next species from the majority of *Lissonota*, although adequate comparative material needs to be accumulated in order to appreciate this character adequately.

Distribution, abundance and phenology: Widespread, although uncommon in occurrence. The species has been recorded as far north as Bridge of Allan, Edinburgh and Morrone Birkwood, Aberdeenshire in Scotland. Ireland (O'Connor *et al.*, 2007): Wicklow. Taken during July to September.

Biology: Nothing is known of the biology of *L. clypealis*, other than an apparent connection with deciduous woodland.

Lissonota gracilipes Thomson, 1877

Taxonomy: Easily distinguished from the preceding species from the characters given – although no males have yet been encountered for *clypealis*. The fore and middle coxae are testaceous to yellow in both sexes.

Distribution, abundance and phenology: Apparently a rare species, which might best be sought via host rearing.

England: Devon, Hembury Wood, ex *Ypsolopha parenthesella* (L.) */ ustella (Clerck)*, coll. 6.vii.; Kent, Dartford Heath, ex *Ypsolopha horridella* (Treits.), *Malus*, 11.vi.; London, Hampstead Heath, ex *Y. horridella* on *Prunus spinosa*, coll. 5.v. [NMS]. Buckinghamshire, Milton Keynes, Linford Wood, ex *Ypsolopha* on *Quercus*, coll. vi.; several, Cambridgeshire, Chippenham Fen, reared from *Y. horridella*, hosts collected vi.2010 (JPB). Flight period: unknown, only reared material has been encountered.

Biology: The usual host of this species appears to be *Ypsolopha*, the biotope being hedgerow and woodland.

Key to species for *saturator* group

The *saturator* group contains the smallest banchines. Those species associated with case-bearing lepidopterous larvae are probably genuinely related, while others are placed here largely as a matter of taxonomic convenience. There could be an element of confusion with smaller members of the *gracilenta* group, some of which also have a transverse depression on tergite 2. However, the depression in question lacks any trace of striation, and the ovipositor is not expanded towards the apex. With regard to one or two species in the *buccator* group which may overlap with regard to possessing a striate depression on tergite 2 – these can usually be recognised on greater size alone (in any case, ovipositor structure can be relied upon as a key character).

Members of this group are very poorly represented in the majority of collections, rarely collected other than via host rearing or in Malaise traps. Males of several species are as yet unknown – which fact makes for exceptional difficulty in attempting to provide realistic guidance in the identification of that sex.

1. Distal flagellar segments with ventral punctate sensilla in female ....
    .... cf. ***nigridens*** Thomson [*buccator* group, p. 100]

–   Distal flagellar segments lacking punctate sensilla 2

2. Antenna pale testaceous in first half (over 8 or more segments); interocellar distance less than 0.5 x ocellus to occipital carina; area of speculum punctate (Fig. 215); metasoma predominantly dark. *Rare* ....
    .... ***antennalis*** Thomson (p. 86)

–   Antenna at most with a few basal segments reddish (or more or less entirely *dark* testaceous); interocellar distance distinctly greater than 0.5 x ocellus to occipital carina; speculum coriaceous or shining (or if absent, metasoma extensively reddish) 3

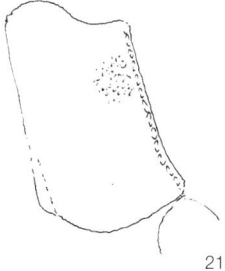

215

3. Mesonotum extensively reddish with yellow and black patterning (metasoma red, marked with black); ovipositor strongly explanate towards apex, very wide in proximal section, and approximately straight to distinctly *down*-curved distally (Fig. 216). *Uncommon* .... ***luffiator*** Aubert (p. 86)

–   Mesonotum black, at most with small yellowish shoulder marks; if ovipositor strongly explanate, then much narrower in proximal section, and approximately straight to distinctly *up*-curved (cf. Figs 219, 222) 4

4. Posterior region of tergite 2 striate for around 0.5 x its length (Fig. 217); interocellar distance only about 1.25 x ocellus to eye; (ovipositor barely explanate towards apex; metasoma with tergite 1 red apically, tergites 2-3/4 reddish, second with piceous markings on dorsum). *Uncommon* .... ***stigmator*** Aubert (p. 86)

–   Zone of striation in trans-impression of tergite 2 more restricted in area (cf. Figs 221, 226); interocellar distance at least equal to 1.5 x ocellus to eye, excepting when metasoma dark with reddish posterior tergal bands; (ovipositor often explanate towards apex; metasoma often predominantly dark in colour) 5

5. Central tergites red, with tergite 3 around 1.4 to 2 x wider than long; first tergite with heavy puncturation and deep striation, second tergite almost rugose-punctate, plus with the striae tending to run diagonally; speculum large, glabrous and shining; (minimum genal length nearly 0.5 x eye length; ovipositor up-curved, barely explanate). *Rare* .... ***fletcheri*** Bridgman (p. 87)

–   Central tergites dark, at most suffused or margined with reddish testaceous (if extensively reddish – then more elongate and with quite different sculpturing; speculum much reduced; minimum genal length often much less than 0.5 x eye length; ovipositor various) 6

6. Flagellum 2 at least 4 x longer than broad; flagellum 1 from 4.5 to 5 x longer than broad (Fig. 218); ovipositor at least equal to metasoma plus propodeum, only weakly explanate at apex (Fig. 219); minimum genal length greater than 1.5 x malar space. *Not uncommon; hosts leaf-rollers* .... ***mutator*** Aubert (p.87)

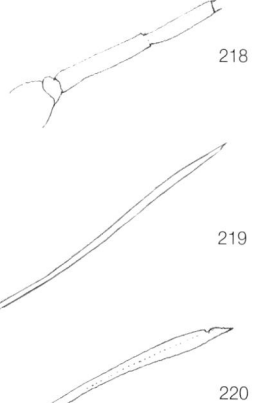

–   Flagellum 2 not more than 3 x longer than broad; flagellum 1 at most 4.5 x longer than broad (if ovipositor greater than or equal to metasoma plus propodeum – then distinctly explanate (cf. Fig. 220); minimum genal length sometimes less than 1.5 x malar space). *No common species; hosts case-bearing incurvariids or psychids* 7

7. Postpetiole with the pre-apical depression forming a deep groove traversed by strong longitudinal striation – tergite 2 with the transverse depression distinct, and with well-developed striation (Fig. 221); (ovipositor distinctly expanded towards apex (Fig. 222); central tergites blackish, with narrow reddish incisures). *Rare* ....
.... ***consobrina* sp. nov.** (p. 87)

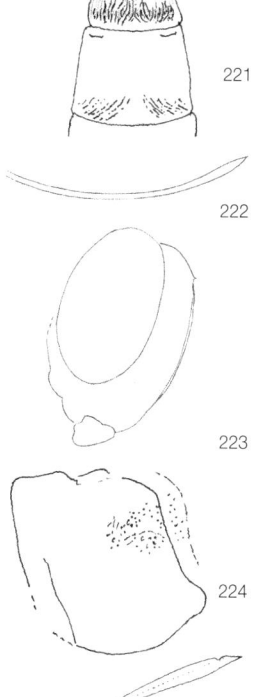

221

222

– Postpetiole with the pre-apical depression relatively shallow, the striae weak to moderately strong; tergite 2 with the depression weak, and with striation absent to weak; (ovipositor often only weakly expanded towards apex central tergites often of reddish or testaceous ground colour) 8

8. Genae distinctly expanded ventrally (Fig. 223); speculum almost occluded by punctures (Fig. 224); ovipositor distinctly expanded apically (Fig. 225); central tergites reddish, with areas of piceous suffusion on dorsum; (tergite 1 with much reduced striation; areolet pentagonal). *Rare* .... ***obsoleta* Bridgman** (p. 89)

223

224

225

– Gena not widening below; speculum prominent; ovipositor at most weakly expanding towards apex; central tergites dark testaceous to piceous in ground colour 9

9. Distance between hind ocelli at most 1.3 x posterior ocellus to eye (Fig. 227); areolet broadly pentagonal; flagellar segment 1 about 4.5 x longer than broad; mesosulcus not widening towards apex (Fig. 228); lateral striation of tergite 1 weak to absent (Fig. 226); (central tergites piceous with distinctive reddish apical margins – which latter may be broad). *Rare* ....
.... ***virgata* sp. nov.** (p. 88)

– Distance between hind ocelli at least 1.3 x posterior ocellus to eye (Fig. 229); areolet quadrangular; first segment of flagellum not more than 4 x longer than broad; mesosulcus widening towards apex (Fig. 230); tergite 1 with lateral striae strongly defined (Fig. 231); (central tergites dark testaceous, with greater or lesser degree of piceous suffusion). *Widespread* .... ***saturator* (Thunberg)** [*basalis*] (p. 90)

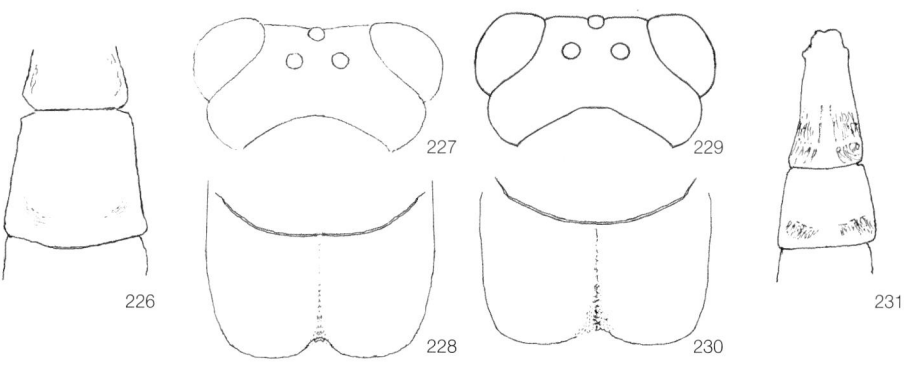

226   227   229   231

228   230

Species accounts for *saturator* group

Lissonota antennalis Thomson, 1877

Taxonomy: This species is easily recognisable from the antennal colouration alone, which is unique within *Lissonota*.

Distribution, abundance and phenology: The only British material was collected in East Anglia, and in fenland in north Wales.

England: Cambridgeshire, Chippenham Fen, carr at reed bed edge, Malaise trap, 26.vii.-10.viii.; Norfolk, Catfield, mature fen, Malaise trap, 6-16.viii.; Norfolk, Santon Downham, 20.vii.-11.viii. (*Betula*, *Pinus*); N. Wales: Barmouth, on *Quercus* [NMS].

Biology: Little is known of the biology of the species, apart from an apparent link to fenland habitats. The host is unknown.

Lissonota luffiator Aubert, 1969

Taxonomy: The colour pattern of mesonotum and metasoma render this species distinctive.

Distribution, abundance and phenology: *L. luffiator* is uncommon, although collectability may well be affected by greater ease of rearing from the host.

England: Kent, Gt. Chatteris Wood, ex *Luffia fershaultella* (Steph.) – coll. 5.iii.; Berkshire, Silwood Park, Malaise trap, vii.; Hertfordshire, Bricket Wood, ex *Psyche* (*Proutia*) *betulina* (Zell.) [NMS]; Huntingdonshire, Monks Wood. vii. [JPB]. Flight period: July (following the limited data available).

Biology: The hosts are Psychidae, feeding on tree trunks and similar locations.

Lissonota stigmator Aubert, 1972

Taxonomy: Another *Lissonota* that is not difficult to distinguish from related species, on the basis of distinctive sculpturing and colour pattern.

Distribution, abundance and phenology: The few available records are quite widespread, and the species may ultimately prove to be co-extensive with that of the host. The species is of Holarctic distribution (Townes and Townes, 1978).

One female: England: Berkshire, Reading University, ex *Anthophila*, coll. 12.v.; one female, Oxford Canal, also ex *Anthophila*, coll. 22.v.; one male, Berkshire, Aylesbury, Weston Turville NNR, (same host), coll. 23.v.; wild-caught adults: Scotland: one male, Tayside, Paddockmuir Wood, Malaise trap: 23-iii.-6.iv.1988; also one female, same location, 18.iv.1987 [NMS].

Biology: The host of *L. stigmator* is *Anthophila fabriciana* (L.) feeding on *Urtica*.

Lissonota fletcheri Bridgman, 1882

Taxonomy: *L. fletcheri* is quite probably misplaced in the present species group. The species is distinctive within *Lissonota* as a whole. Males lack the transverse impression on tergite 2, but otherwise agree with the female diagnosis.

Distribution, abundance and phenology: *L. fletcheri* is little known in the UK, and there are no recent records. Its distribution extends to Lancashire and Yorkshire.

England: TYPES: lectotype female: Worcestershire, paratype male, same data; Nottinghamshire, Crow Wood Hill, West Leake (Morley); Merseyside, Southport 'ex *Tortrix* sp.' (Lyle) [NHM]. One female, England: 1877 'Wor'ch', 'ex larva of ?*Gelechia lentiginosella*' (Cameron) [NMS]. Yorkshire, Potteric Carr, 3.vi.1982 [WE]. Ireland (O'Connor *et al.*, 2007): Donegal, amongst *Salix* (teste Johnson – unconfirmed). Flight period: June (according to the fragmentary data available).

Biology: The host is reputedly *Mirificarma lentiginosella* (Zell.) on *Genista*. However, another source indicates *Depressaria hypericella* (*Agonopterix liturosa*). [NHM]. The two reputed host species probably conceal at least one error. Thus it is not possible to draw any definite connection concerning biotope.

Lissonota mutator Aubert, 1969

Taxonomy: Somewhat isolated within the present species group on account of the gregarious larval stage. Characters of the species group are ill-defined in males of *mutator*.

Distribution, abundance and phenology: *L. mutator* appears to be widespread – although seldom collected other than via Malaise trapping or by rearing from hosts.

England: one female, two males, Hampshire, Pamber Forest, cocoons on *Prunus*, coll. 12.vii.; three females, Hampshire, Pamber Forest, cocoons in folded *Corylus* leaf; two females, Cumbria, Beetham, ex tortricid on *Prunus*, coll. 21.v. [NMS]. Surrey, Ashtead Common, oak with willow scrub, Malaise trap, 18.vi.-6.vii.; also Chobham Common and Windsor Forest [HM].

Biology: *L. mutator* is a gregarious parasitoid of *Tortrix viridana* (apparently also of other arboreal Lepidoptera), usually linked to deciduous woodland.

Lissonota consobrina sp. nov.

Female: flagellar segment 1 from 3.6-3.8 longer than broad. Temples narrowing, somewhat convex in dorsal aspect; occipital carina weakly angled. Sculpture of temple: finely coriaceous, obscurely punctured. Frons finely coriaceous, distinctly punctate. Interocellar distance at least 1.5-1.6 x posterior ocellus to eye, and 0.66 to almost equal to distance to occipital carina. Maximum temple length v. flagellum 1 length: about 0.65-0.75. Minimum genal length about 1.5 x malar space, and about 0.8 x basal width of mandible. Fore wing length 4.5-5 mm. Mesopleurum weakly coriaceous, punctures approximately equal to interspaces, becoming larger above speculum – latter extending about 0.4 x distance to epicnemium. Mesosternal sulcus with the transverse costae strong and uniform throughout. Propodeum: dorsal longitudinal carinae well-defined; area basalis weakly convergent – sculpture: punctures distinct, approximately same diameter as interspaces, becoming more or less rugose towards posterior. Metasoma: tergite 1 from 1.5-1.6 x longer than broad, coriaceous medially, laterally with conspicuous striae, pre-apical depression forming a deep groove traversed by strong striation. Tergite 2 broader than long, coriaceous, pre-apical depression striate. Ovipositor longer than metasoma, shorter than metasoma plus propodeum. Colour: black, metasoma with central tergites having narrow pale margins; legs entirely testaceous.

Male: From female: mesonotum with yellow shoulder marks; interocellar distance greater than posterior ocellus to occipital carina; mesosternal costae nearly absent.

HOLOTYPE. ENGLAND: female, Essex, Stanford-le-Hope, ex *Nemophora fasciella* (F.) on *Ballota nigra*, coll. 18.iii.1989, em. v.1989 [NMS].

PARATYPES. ENGLAND: one female, one male, same data as holotype; female, Essex, Mucking Cr., ex *N. fasciella* on *Ballota nigra*, case coll. 31.ii., em. vi.1976 [NMS]; two females: Kent, Welmington, 'exc. 2.VIII.35. ex Nemophora fasciella Fletcher'; one male same data, one male same data – except emerged: 7.VIII.1935. One female: 'Bexley Kent em. 7-7-1936, ex Nemophora fasciella. H.W. Daltry', male same data except 9-7-1936. [NHM]. **SWITZERLAND**: one female, Suisse-Vaud '+ erryres', 8.ix.1964. (Beaumont) [Lausanne Museum].

Taxonomy: The deep transverse impression on tergite 1 is distinctive (the same trait may be found on some *mutator*, although this problem is readily overcome on account of the different ovipositor structure).

L. consobrina ('cousin') is closely related to *L. rufitarsis* Szépligeti, from which it differs in the sculpture of the postpetiole as described here (also in aspects of the fore wing venation, plus dimensions of the gena and ovipositor).

Distribution, abundance and phenology: In the UK, known only from type material. Flight period: no wild-caught examples have been encountered.

Biology: The host is the case-bearing incurvariid *Nemophora fasciella*. The latter is a very local species, linked to a range of open habitats – including marshland. The related *L. rufitarsis* carries an unconfirmed record from *Nemophora unicincta* (Aubert, 1978).

Lissonota virgata sp. nov.

Female: Flagellum 1 about 4.5 x longer than broad. Temples narrowing in dorsal aspect; occipital carina distinctly angled centrally; interocellar distance 1.2 to 1.33 times posterior ocellus to eye, and about 0.6 x posterior ocellus to occipital carina. Maximum temple length about 0.75 x flagellum 1. Minimum genal length at most 1.25 x malar space. Malar space versus basal width of mandible from approximately equal, to less than 1.3. Fore wing length about 4.5 mm. Mesopleurum coriaceous, with distinct puncturation; speculum extending at most 0.4 x distance to epicnemium. Areolet broadly pentagonal. Propodeum: longitudinal carinae extending a little beyond region of area basalis, weakly divergent. Metasoma: tergite 1 about 1.6 x longer than broad, with subapical depression at most weakly defined, striae weak or absent. Tergite 2: slightly broader than long – pre-apical depression distinct, with moderate striation. Ovipositor longer than metasoma, slightly widening towards apex. Colour: head and thorax black; metasoma piceous, with broad reddish apical margins on central tergites; legs uniform testaceous, trochanters and trochantelli to some extent paler.

Male: unknown.

HOLOTYPE. ENGLAND: female, Berkshire, Windsor Forest, Malaise trap, 26.vi.-27.vii.1992 [NMS].

PARATYPES. ENGLAND: female, same data as preceding; female, Norfolk, Catfield, Malaise trap, abandoned wet meadow, 24.vii.-1.viii.1993 (Jarvis); female, Cambridgeshire, Chippenham Fen, Malaise trap: carr at reed bed edge, 16.-24.vi.1983 (Field); female, Oxfordshire, Taynton Fen, Malaise trap, 16.viii.-8.ix.1989 (Porter) [NMS]; female: Surrey, Ashtead Common, Malaise trap, willow scrub 20.vii.-5.viii.94 (HM).

Taxonomy: The metasomal patterning of the present species is distinctive (the name *virgata* refers to this annulated patterning). Note that *L. picticoxis* (p. 91) has a similar colour pattern and general habitus to *L. virgata*. However, the two species are readily separable on the structural characters given in the key.

Distribution, abundance and phenology: Known only from the type series. Flight period: June to September.

Biology: Hosts unknown. The habitat includes ancient fenland.

Lissonota obsoleta Bridgman, 1889

Female: Flagellar segment 1 from 4, to nearly 5 x longer than broad. Temples narrowing behind eyes in dorsal aspect; occipital carina distinctly angled centrally. Sculpture of head: temple coriaceous, with faint puncturation; frons coriaceous, punctate laterally. Interocellar distance v. posterior ocellus – eye: approximately equal to, to about x 1.25. Maximum temple width about 0.66 x flagellum 1. Minimum genal length 1.3-1.5 x malar space. Malar space distinctly greater than basal width of mandible. Fore wing length about 3.5 mm. Mesopleurum coriaceous with distinct punctures, the latter greater than interspaces; speculum mostly obscured by large punctures. Mesosternal sulcus widening towards apex, trans-costate, closed by several costae. Propodeum with the dorsal longitudinal carinae distinct, area basalis absent. Sculpture: punctate in front, rugose behind. Metasoma: tergite 1 a little less, to a little greater than 1.5 x longer than broad. Sculpture: coriaceous, obscurely punctate laterally from about half length; pre-apical impression weak, with indistinct striation. Tergite 2: coriaceous, trans-impression visible laterally, with or without weak striation; the tergite slightly, to distinctly longer than broad. Ovipositor: a little shorter, to a little longer than metasoma. Colour: black – metasoma with tergites 1-3 reddish, with dark dorsal markings; legs entirely red.

Male: unknown.

NEOTYPE. England: female, Norfolk, Winterton, dunes, Malaise trap, vii.2004 [NCM].

OTHER MATERIAL. England: female, Great Hockham, 4.viii. [NMS]; one female, Surrey, Thursley Common, birch / heath, Malaise trap, 21.vii.-9.viii.; one female Surrey, Chobham Common, scrub, Malaise trap, 28.vi.-25.vii.; one female, Headley Warren, Orchid Bank, Leatherhead, Malaise trap, 1-30.vii. [HM]; female: Huntingdonshire, Monkswood, Malaise trap, Ash-Elm wood, 13-25.vii.2005 [NHM].

Taxonomy: The original type specimens of Bridgman's *L. obsoleta* (one male, one female, West Sussex) have been lost.

Distribution, abundance and phenology: Of rare occurrence in Britain. Existing records are from southern England. Flight period: June to July.

Biology: Possibly linked to sandy and chalky habitats. Host: Psychidae – '? *Bruandia comitella*' on the European continent, from material loaned by Klaus Horstmann.

Lissonota saturator (Thunberg, 1882) [*basalis*]

Taxonomy: The fore and mid coxae vary from testaceous to yellow in both sexes. The species group traits of tergites 1 and 2 are obscured in the male sex – and some convergence with males of *L. gracilenta* is also in evidence. The species is known under the name *basalis* in the older literature.

Distribution, abundance and phenology: Widespread throughout England and Scotland, but uncommon.

England: Cheshire, Abbott's Moss, *Quercus / Betula / Pinus*, Malaise trap; Norfolk, Santon Downham, heath with *Betula* and *Pinus*, Malaise trap [NMS]; Berkshire, Windsor, Highstanding Hill, *Quercus / Juncus*, Malaise trap [HM]; Scotland: Stirlingshire, Flanders Moss, bog / *Betula* / heath, Malaise trap [HM]; Argyllshire, Taynish [NMS]; Wales: Wrexham, Glyn Ceiriog, river valley with *Alnus*, *Acer*, etc, Malaise trap [JPB]. Ireland (O'Connor *et al.*, 2007): Armagh (Johnson). The latter record requires confirmation. Flight period: late May to August.

Biology: *L. saturator* has been reared from the psychid moth: *Diplodoma laichartingella* (Goeze) (= *herminata*) (Geoffr.) [Bland, NMS]. The biotope seems usually to be woodland or heath.

Key to species for *gracilenta* group

There may occasionally be confusion between smaller species in the present array and certain members of the *saturator* group. However, the latter generally have the ovipositor expanding towards the apex, and the transverse depression of tergite two more or less striate.

Sexual dimorphism can be quite pronounced amongst members of the species group. Nevertheless, the majority of known males will run in the key – providing it is recognised that sculpturation of the basal metasomal tergites tends to manifest some degree of pucturation, while being predominantly coriaceous in females.

1. Tergite 1 at least 2 x longer than broad; tergite 2 with some scattered puncturation towards basal half; tergite 3 distinctly longer than broad (Fig. 232); flagellum with at least 34 segments; fore wing length about 6 mm; (scutellum and notum 2 often reddish-marked, mesopleurum rarely reddish-suffused); fore wing length about 5 mm. *Common* .... ***tenerrima*** Thomson (= *variabilis*) (p. 91)

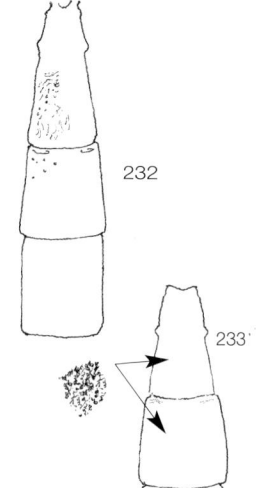

– Tergite 1 less than 2 x longer than broad; tergite 2 lacking puncturation (Fig. 233); tergite 3 transverse to subquadrate; flagellum with 31-32 segments; fore wing length not greater than 5 mm; (if scutellum and / or notum 2 pale-marked, then mesopleurum is likewise); fore wing length at most 4 mm 2

2. Thorax with at least mesopleurum and scutellum richly red-marked 3

– Thorax black 4

3. Areolet 4-sided; minimum genal length only about 0.25 x eye length; tergite 1 uniformly coriaceous, with some striae laterally towards apex, length greater than 1.5 x apical width (Fig. 234); tergite 2 subquadrate; laterotergite 4 inflexed at base alone (Fig. 235); mesonotum black, only scutellum, pro- and mesopleurum, also propodeum laterally, red-marked; fore and mid coxae yellow, tibial bases red). *Rare* .... *picticoxis* Schmiedeknecht (p. 91)

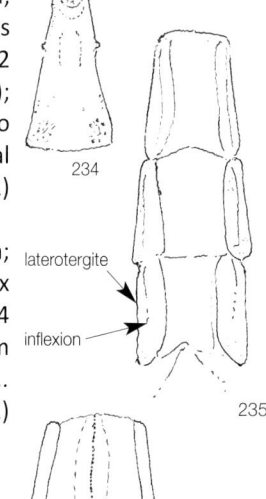

–   Areolet pentagonal; minimum genal length about 0.4 x eye width; tergite 1 with faint diagonal microsculpture, length at most 1.4 x apical width; tergite 2 about 1.3 x broader than long; laterotergite 4 at least half inflexed (Fig. 236); mesonotum (also pleura, scutellum and coxae) entirely red, tibial bases yellow. *Little known in the UK* .... *halidayi* Holmgren (p. 92)

4.. Temples weakly convex and narrowing in dorsal view (Fig. 237); minimum genal length not greater than length of flagellum 4, and about 1.5 x malar space; speculum coriaceous; fore coxae frequently yellow-marked (entirely yellow in males). *Common* .... *gracilenta* Holmgren (p. 92)

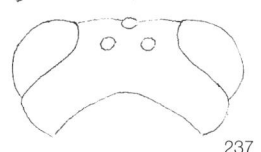

–   Speculum at least partly shining (temples often distinctly convex) .... cf. *carbonaria/simulator* [*buccator* group] (p. 102)

Species accounts for *gracilenta* group

Lissonota tenerrima Thomson, 1877 (*variabilis* syn nov.)

Taxonomy: Males have the characteristically elongate first tergite as in the female sex. In terms of structure, some examples of both sexes with more than the usual degree of tergal puncturation may converge upon the *buccator* group (e.g. if the speculum is shining and tergite 2 is extensively punctate – cf. notes on *L. palpator*, p. 101). In the past, *L. tenerrima* has been split on the basis of colour patterning, including *variabilis* – here interpreted as a new synonym, on the basis of an analysis of over twenty structural characters.

Distribution, abundance and phenology: Generally distributed and common, at least to Argyll and Inverness in Scotland. Ireland (O'Connor *et al.*, 2007): Armagh; Down (as *variabilis*). Flight period: during May to August.

Biology: The species attacks larvae of *Scoparia* (sensu lato) living in mosses on tree trunks, walls, etc.. There are verified rearings from: *Dipleurina lacustrata* (Panzer), *Eudonia truncicolella* (Staint.), also *Aplota palpella* (Haw.) in *Hypnum.* (NMS). There are confirmed records from M.V. light traps.

Lissonota picticoxis Schmiedeknecht, 1900

Taxonomy: A very distinctive species, at once recognisable on the basis of colour characteristics (although, see comments on *L. virgata*, p. 88).

Distribution, abundance and phenology: *L. picticoxis* is little known in the UK, having been recorded from a few specimens only.

England: Yorkshire, Malham Tarn, 20.viii.1955, det. J. F. Perkins [MU]. Scotland: Sutherland, *Pinus sylvestris* plantation on blanket bog, Malaise trap, 1-13.viii. [NMS].

Biology: One example has been reared from *Infurcitinea albicomella* (Herr.-Schaeff.), on *Cotoneaster microphyllus* (presumably from a fungus growing on this plant), Devon, Torquay, 26.iv.1991, Heckford [NMS].

Lissonota halidayi Holmgren, 1860

Taxonomy: *L. halidayi* is another very distinctive species with regard to colour pattern.

Distribution, abundance and phenology: Little known in the UK.

England: one female, Boxmoor, 1934 (Benson) [NHM]. Flight period: unknown.

Biology: Unknown (Morley's reference to reared material was based on a misdetermination).

Lissonota gracilenta Holmgren, 1860

Taxonomy: *L. gracilenta* resembles a smaller, broader form of *tenerrima*. However, the wide metasoma is a good recognition character. Some degree of sexual dimorphism is in evidence. Males have similar metasomal sculpture to females, except that there is some striation on tergite 1, and a little puncturation on the second tergite; the face, frons and thorax have a varying degree of yellow patterning, the front and middle coxae are yellow. The central tergites have narrowly testaceous apical margins in females – more broadly so in males.

Distribution, abundance and phenology: Despite being poorly represented in older collections, *L. gracilenta* appears now to be widespread and not uncommon, occurring at least as far north as Perthshire in Scotland. Flight period: June to August.

Biology: Aubert (1978) gives *Grapholita molesta* (Busck) as a reputed host (f. Pisica, 1973). This reference requires confirmation.

Key to species for *versicolor* group

All species have profuse reddish areas on the thorax. Two members of the group have been reared via scopariine Pyraloids, which may indicate that the assemblage is a natural one; fore wing lengths 4-6 mm.

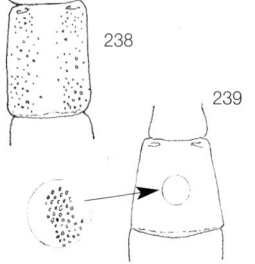

1. Central tergites with restricted zones of puncturation (Fig. 238), i.e. more or less impunctate in median zone; (flagellum 1 from 4-4.5 x longer than broad). *Not uncommon* .... **culiciformis** Grav. (p. 93)

– Central tergites strongly and closely punctured – including median area (Fig. 239); (flagellum 1 at least 5 x longer than broad) 2

2. Flagellum 1 at least 6 x longer than broad (from 5 x in males) (Fig. 240); mesonotum profusely red-marked, apices of coxae and trochanters yellow-marked (entirely yellow in males). *Uncommon* .... ***versicolor*** Holmgren (below)

240

– Flagellum 1 about 5 x longer than broad (Fig. 241); mesonotum black, legs entirely red (male unknown). *Rare* .... ***pleuralis*** Brischke [*strigifrons*] (below)

241

Species accounts for ***versicolor*** group

Lissonota culiciformis Gravenhorst, 1829

Taxonomy: The combination of colour pattern and tergal sculpturing render this species easily recognisable – although care should be taken to avoid confusion with some colour forms of *L. tenerrima* [*gracilenta* group].

Distribution, abundance and phenology: Widely distributed, uncommon; existing records extend to Wales. Flight period: July and August.

Biology: Habitats include woodland and suburban gardens. Aubert (1978) lists some (unconfirmed) hosts.

Lissonota versicolor Holmgren, 1860

Taxonomy: Another easily recognisable species on the basis of its distinctive colour pattern, plus structural traits given in the key.

Distribution, abundance and phenology: Uncommon, and apparently restricted to southern counties, so far as the UK mainland is concerned.

England: Oxfordshire, Wychwood Forest, Malaise trap; Devon, Bolthead, ex *Eudonia lineola* (Curt.), under *Xanthoria parietina* on rock; same host / host pabulum: Cornwall, Kynance Cove. Ireland: Arran Islands, Inishmore [NMS]; Kerry, Blasket Island [NHM]; (O'Connor *et al.*, 2007): Antrim. Flight period: July and August.

Biology: *L. versicolor* has been reared from a single scopariine species on several occasions (see above data) – which infers an oviposition territory amongst mosses growing upon tree trunks and similar substrates.

Lissonota pleuralis Brischke, 1880 [*strigifrons*]

Taxonomy: A distinctive species on the basis of colour patterning.

Distribution, abundance and phenology: Scattered records from southern England only.

England: 26 females, Norfolk, Santon Downham [NMS]. Female, Buckingham Palace gardens, Malaise trap, 15.viii.96 [NHM]. Flight period: late July to late August.

Biology: *L. pleuralis* has been reared from *Scoparia lineola* (Curt.): Isle of Purbeck (Morley coll.). [NHM]. The habitat is obviously linked to that of pyralid larvae living in mosses. The Santon Downham data (see above) imply possible thelytoky.

Species account for *dubia* group

Lissonota dubia Holmgren, 1856

Taxonomy: It should be noted that the type male of *L. dubia* is missing, and identification follows other Holmgren material preserved under the same name. *L. dubia* (as recognised here) can usually be distinguished from other small *Lissonota* species on the basis of the sculpture of propodeum and of tergite 1. In the event of possible confusion with the *saturator* group (and in particular, with *L. saturator* itself), the lack of any transverse impression on tergite 2, plus absence of any distal widening of the ovipositor will generally suffice for avoidance of confusion. The only other small *Lissonota* with which *dubia* might possibly be confused is *L. nigridens* – which has distinctive sensilla on the distal segments of the antennal flagellum and heavier puncturation. In both instances, the link between the present species and coniferous woodland will act as an additional guide to correct identification.

Distribution, abundance and phenology: Widespread, although poorly represented in collections; possibly more frequent in the Scottish Highlands.

England: two females, Suffolk, Brandon (Morley)**;** male, female: from *Epinotia tedella*, Forestry Commission [NHM]. Scotland: Inverness-shire, Loch Garten, *Pinus* and *Betula*, Malaise trap; Rothiemurchus, Cairngorms NNR, native pine, Malaise trap, 1-13.v.; Glen Clova, ex *Epinotia tedella* in *Picea* shoots, coll. 8.xi.; Perthshire, Coir Choille Chuilc, native pine, Malaise trap, vi. [NMS]. Ireland (O'Connor *et al.*, 2007): Donegal; Down; Mayo. Flight period: May to October (indicating a probable two-generations life cycle).

Biology: The species is associated with coniferous forests, where the known host *Epinotia tedella* (Clerck) occurs. Given the very broad distribution of the host species, it seems surprising that *L. dubia* has not been taken more widely in Britain.

Key to species for *buccator* group

The *buccator* group is the largest sub-section of the genus *Lissonota*. *L. nigridens* is easily recognised (at least in the female sex), as are those species carrying transverse basal metasomal tergites. Those with elongate tergite 1 and central tergites also create no real difficulty in identification. The main problem areas lie with two species complexes centred about *L. carbonaria*, and *buccator* itself. In order to successfully identify members of these two groups, it is necessary to form at least a modest collection, in order that meaningful comparisons can be made.

The *buccator* and *carbonaria* sections are probably natural units. Other members are placed here as a matter of convenience.

Buccator group fore wing lengths fall into the range 3.5–5.5 mm, usually tending towards the higher end.

1. Metasoma with extensive red markings 2

–   Metasoma with at most the incisures reddish 4

2. Central tergites strongly transverse, at least 1.35 x broader than long; temples a little more convex in dorsal aspect (Fig. 242); tibia 3 not paler at base; (minimum genal length about 2 x width of malar space). *Rare* .... *maculata* Brischke (p. 99)

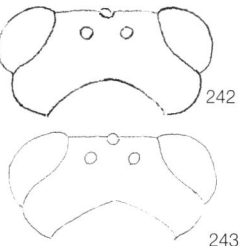
242
243

— Central tergites at most slightly broader than long; temple less convex in dorsal aspect (Fig. 243); (tibia 3 sometimes white at base) 3

3. Flagellum 1 at least 5 x longer than broad; tergites 2-3 subquadrate to longer than broad, usually to a greater or lesser extent suffused with dark pigmentation; hind tibia entirely red. *On dead timber, often common* .... *semirufa* (Desvignes) (p. 99)

— Flagellum 1 at most 4 x longer than broad; tergites 2-3 subquadrate to a little broader than long, usually predominantly reddish; hind tibia white basally. *Not uncommon* .... *quadrinotata* Grav. (p. 100)

4. Distal flagellar segments with ventral punctiform sensilla in female sex (Fig. 244); genal carina joining hypostomal near base of mandible (as Fig. 245); areolet open – or with outer vein less than half complete; (occipital carina more or less angled (Fig. 246); trochanters usually darkened). *Common* .... *nigridens* Thomson (p. 100)

— Distal flagellar segments without ventral punctiform sensilla; genal carina almost always joining hypostomal distinctly behind base of mandible (as Fig. 247); areolet with outer vein generally at least half complete; (occipital carina often rounded; trochanters usually not darkened) 5

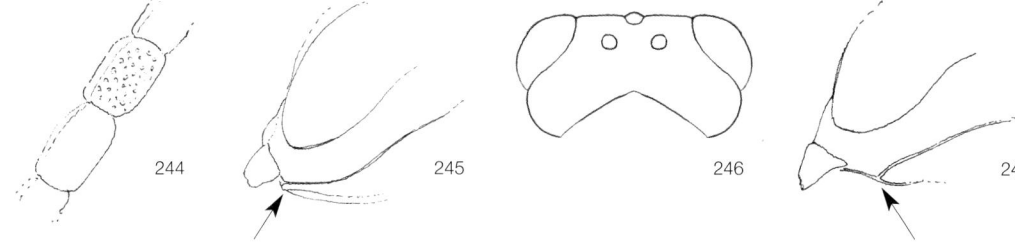
244   245   246   247

5. Flagellar segment 1 around 5 x longer than broad; head with a continuous yellow-line along facial orbits, continuing over frons to vertex (sometimes narrowly interrupted); scutellum and often coxae yellow-marked; (speculum 'closed' by zone of puncturation (Fig. 248); central tergites subquadrate to longer than broad). *Common* .... *palpalis* Thomson (p. 100)

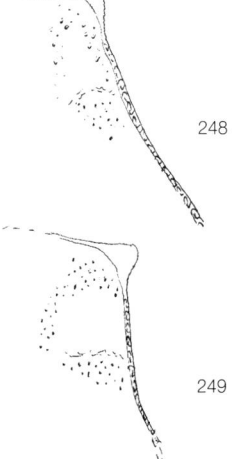
248
249

— If rarely, flagellar segment 1 is greater than 5 x longer than broad – frontal and vertical orbits not continuously yellow-lined (if frontal orbit yellow-margined, line does not continue to vertex, and speculum is 'open' – continuing more or less uninterrupted to posterior margin of mesopleurum) (Fig. 249); scutellum, and usually coxae, lacking yellow markings 6

6. Tergite 1 not less than 1.7 x longer than broad; central tergites subquadrate to distinctly longer than broad (often distinctly punctate, at least in anterior portion (Figs 250, 251)) 7

−   Tergite 1 rarely more than 1.6 x longer than broad; central tergites subquadrate to strongly transverse − sometimes indistinctly punctate, and then with coxae and trochanters dark piceous 10

7. Interocellar distance at most 0.4 x ocellus to occipital carina, and not more than 0.66 x ocellus to eye (tergite 2 with large punctures, mostly wider than interspaces − the central tergites strongly coriaceous; ovipositor shorter than metasoma; hind trochanters and trochantelli infuscate; flagellar 2 about 2.5 x longer than broad). *Little known in the UK* .... **trochanterator** Aubert (p. 101)

−   Interocellar distance more than 0.5 x ocellus to occipital carina, and usually not less than ocellus to eye distance (differing in at least one other character) 8

8. Speculum largely coriaceous; tergite 1 predominantly coriaceous, excepting toward apex (males with scutellum often yellow-marked) .... cf. **tenerrima** (p. 91)

−   Speculum at least partly shining, with or without some degree of puncturation; tergite 1 predominantly with deep puncturation 9

9. Central tergites strongly and closely punctate, with at most a very narrow median impunctate zone (Fig. 250); flagellar segment 1 at least 5 x longer than broad, flagellar 2 at least 3.4 x longer than broad. *Not uncommon* .... **palpator** Aubert (p. 101)

−   Central tergites with broad shining glabrous apical margin, and with punctures narrower than interspaces, becoming much sparser in posterior half, and with a wide median impunctate zone (Fig. 251); flagellar segment 1 at most a little greater than 4 x longer than broad, flagellar 2 less than 3 x longer than broad. *Rare* .... **accusator** (F.) [*unicincta*] (p. 102)

10. Tergite 2 virtually impunctate, or with indistinct puncturation (latter mostly smaller than interspaces), amongst predominantly transverse coriaceous microsculpture (e.g. Fig. 252); tergite 1 likewise − at least basad of pre-apical impression (tergite 5 with similar sculpture and vestiture to preceding tergite; tergite 6 uniformly hairy (Fig. 253). [N.B. if speculum entirely coriaceous − cf. *gracilenta* group, p. 91] 11

−   Tergite 2 with distinct puncturation (some proportion of punctures being greater than or equal to interspaces) (e.g. Figs 270, 271); tergite 1 often conspicuously punctate basad of pre-apical impression (tergite 5 often with weak microsculpture, its baso-dorsal region bare; tergite 6 often bare over baso-dorsal region (as Fig. 254)) 13

250

251

252

253

254

11. Temples distinctly convex behind eyes (Fig. 255); if tergite 2 punctate, then punctures at most equal to interspaces; head usually all black – occasionally with small pale marking on vertex and / or short yellowish line far ventrad of antennal scrobes on facial margins; clypeus with unusually dense hair (Fig. 256b), temples sometimes with long lateral pubescence (Fig. 257a); (tergite 1 with apical impression usually only weakly defined, striation generally confined to lateral aspect (see Fig. 252); flagellum 1 from 4.5-5 x longer than broad) 12

– Temples distinctly declivous behind eyes in dorsal aspect (Fig. 258); frontal part of tergite 2 with at least some punctures larger than interspaces (if in doubt – then usually at least head and frontolateral angles of mesonotum with prominent yellowish markings); clypeus with normal pubescence (Fig. 256a), temples always with short lateral pubescence (as Fig. 257b); (tergite 1 with apical impression often with striation across dorsum (Fig. 259)) 13

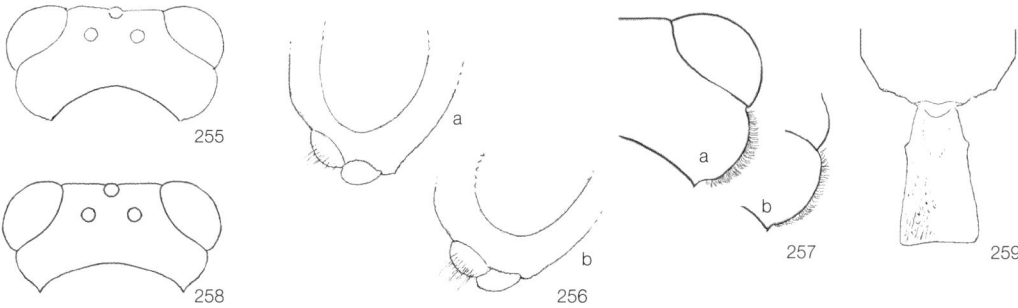

12. Pubescence on lateral temple longer and denser – distinctly longer than diameter of an ocellus (Fig. 257a); speculum smaller – closed by up to 4 rows of punctures. *Hosts in rotten timber* .... *simulator* **sp. nov.** (p. 103)

– Pubescence on lateral temple shorter and sparser, not longer than diameter of an ocellus (Fig. 257b); speculum larger, open – or else closed by up to 2 rows of punctures. *Hosts in plant galls* .... *carbonaria* Holmgren (p. 102)

13. Flagellum 1 usually about 5, to over 6 x longer than broad; apical impression of tergite 1 weakly defined, non-striate on disc, width of postpetiole at least 0.66 x width of propodeal apex (Fig. 260); propodeum more or less impunctate on dorsum; speculum about 0.3 x distance between epicnemium and hind margin of mesopleurum; facial orbits with yellow margin restricted to a small mark adjacent to antennal scrobes (Fig. 261) pronotum black (tergite 2 with weak puncturation; fore wing length about 4.5 mm). *Hosts in rotten timber. Rare* .... *arborator* sp. nov. (p. 104)

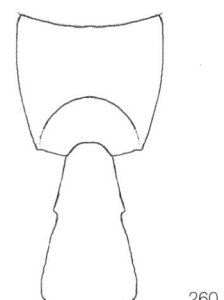

– Flagellum rarely more than 4.5 x longer than broad; tergite 1 with the apical impression often marked with striation across dorsum (see Fig. 259); (width of postpetiole often less than 0.66 x width of propodeal apex); propodeum generally with distinct puncturation on dorsum; speculum about 0.4 x distance between epicnemium and hind margin of mesopleurum; facial orbits often with more extensive yellow margin, tergite 2 often conspicuously punctate, pronotum often yellow-marked 14

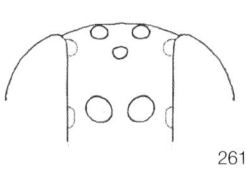

14. Tergite 2 punctures mostly small, not greater than interspaces over median portion, becoming similar to interspaces in frontal region, the puncturation intermixed with conspicuous transverse microsculpture (Fig. 262); postpetiole of tergite 1 less than 0.66 x width of propodeal apex (Fig. 259); nervellus subvertical to weakly obtuse-angled (Fig. 263); internal orbit often with yellow margin of varying length; (pronotum often with prominent yellow markings, subtegular ridge often yellow; all coxae and trochanters reddish; tergite 2 around 1.3 x broader than long). *Abundant in woodland* .... *folii* Thomson [*transversa*] (p. 104)

– Tergite 2 with stronger punctures, usually predominantly larger than interspaces, mostly tending to obscure the underlying microsculpture (see Figs 270, 271) (except rarely, when coxae and trochanters are black); postpetiole at least 0.66 x width of propodeal apex (cf. Fig. 260); nervellus weakly to strongly obtuse-angled (Fig. 264); (face often lacking lateral pale markings; if pronotum with yellow dorsal markings and subtegular ridge yellow – tergite 2 not less than 1.4 x broader than long) 15

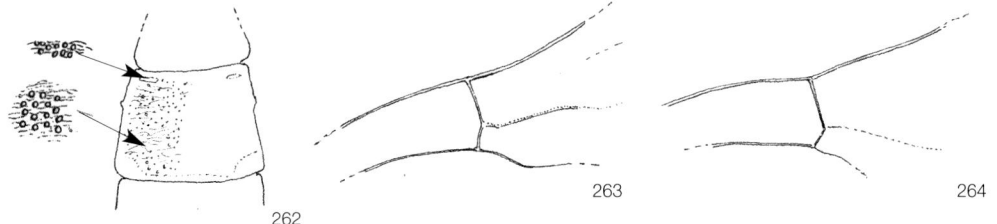

262   263   264

15. Tergite 2 with punctures predominantly narrower than interspaces (Fig. 265); mesosternal sulcus with very heavy transverse costae in posterior half (Fig. 266) ; speculum very small (about 0.25 x distance to epicnemium); all coxae and trochanters dark brownish-black (if reddish testaceous – cf. *accusator*, p. 102). *Little known* ....
.... *serena* sp. nov. (p. 105)

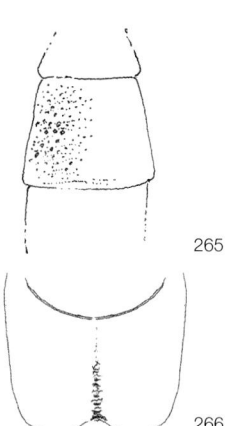

265

266

– Tergite 2 with punctures mostly wider than interspaces (see Figs 270, 271); mesosternal sulcus with relatively weak transverse carinae; speculum larger; coxae and trochanters reddish, sometimes yellow-marked 16

16. Flagellum 1 from 3-3.5 x longer than broad, expanding toward apex (Fig. 267); minimum genal length at least equal to length of third flagellar segment. *Uncommon* .... *punctiventrator* Aubert (p. 106)

267

– Flagellum 1 from 4-4.5 x longer than broad, parallel-sided (e.g. Fig. 268); minimum genal length narrower than length of third segment of flagellum 17

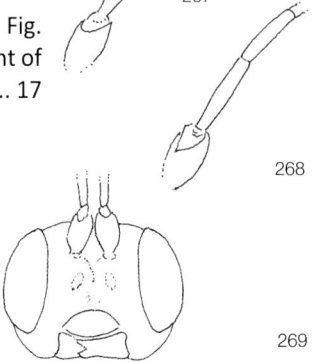

268

269

17. Face usually with reddish mark either side of median line (Fig. 269); fore and mid coxae usually yellow-marked; tergite 2 usually about 1.4 x broader than long, uniformly punctate – tergite 1 with the pre-apical impression shallow, and at most with weak striation over apico-dorsal region (Fig. 270); mesonotum with prominent fronto-lateral yellow markings. *Widespread, not uncommon ..... **buccator** (Thunberg) (p. 106)

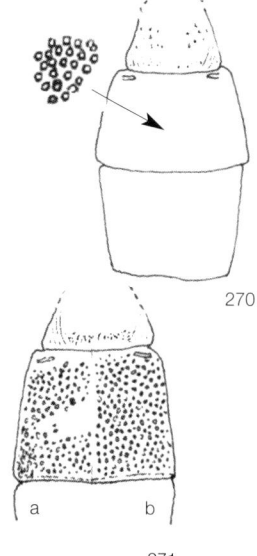

270

– Face without pale markings; fore and mid coxae testaceous, not yellow-marked; tergite 2 often subquadrate, usually not more than 1.3 x broader than long, puncturation often absent in part – tergite 1 generally with a stronger pre-apical impression, latter more often distinctly striate in part, this sculpture extending over apico-dorsal region (Fig. 271a, b); thorax at most with small, roundish shoulder marks. *Widespread, sometimes common* .... **punctiventris** Thomson (p. 107)

a   b

271

Species accounts for *buccator* group

Lissonota maculata Brischke, 1865

Taxonomy: Members of the 'red section' of the *buccator* group are not difficult to identify. However, care must be taken in order to avoid confusion with the *coracina* group (see p. 81), members of which are superficially similar (and much more frequently encountered in the field).

Distribution, abundance and phenology: Little known in the UK.

England: Berkshire, Windsor Forest (Donisthorpe), 1935; S.E. London, Norwood (metasoma missing, specimen det. J. F. Perkins) [NHM]. Ireland (O'Connor *et al.*, 2007): Clare; Dublin; Wicklow (these latter records require confirmation, due to earlier confusion with *L. semirufa* Desv.). Flight period: incomplete data on existing specimens.

Biology: Unknown.

Lissonota semirufa (Desvignes, 1856)

Taxonomy: This is the usual *Lissonota* with reddish marked central tergites that is encountered on dead timber. No male examples of the species have yet been found.

Distribution, abundance and phenology: Widely distributed, although apparently commoner in the south – sometimes occurring in great numbers around standing dead timber. Distribution records extend northwards to West Lothian, Tayside and Sutherland. Flight period: usually to be taken during early June to mid July, sometimes during the latter part of May.

Biology: An 'association with' *Esperia sulphurella* (F.) has been encountered in data received along with material of the species – teste Hickford (on *Pinus*), Bradford, and Langmaid (also: 'could be *geoffrella*' – Bradford). The habitat includes both deciduous and coniferous woodland, with most material collected in the former. There is a possibly valid rearing ex *sulphurella* (teste Lyle), also thirty-two females reared together from Larch logs [NMH]. Other instances of large numbers of females with no males being collected from a single location indicate probable thelytoky.

Lissonota quadrinotata Gravenhorst, 1829

Taxonomy: Apart from taxonomic characters given in the key, the association with the *Lonicera*-feeding *Ypsolopha dentella* (F.) offers a clear guide to recognition of the present species.

Distribution, abundance and phenology: Rare in collections – but probably belonging to that category of species which is most easily encountered through host rearing. On the latter basis (combined with data from Malaise trap material), *L. quadrinotata* is apparently widespread. Reared material from: England: Berkshire; Worcestershire; Lancashire; Cumbria; Buckinghamshire. Malaise trap material from: Norfolk, Catfield; Wiltshire, Savernake Forest; Oxfordshire, Blenheim Park; Hampshire, Ashford Hill Meadow. Wales: Anglesey; Tregarth [NMS]. Flight period: during July and August.

Biology: To be found amongst *Lonicera* growing on deciduous trees.

Lissonota nigridens Thomson, 1889

Taxonomy: Despite the small size and insignificant appearance of the species, it is readily determined on the basis of the unusual morphology of the distal flagellar segments.

Distribution, abundance and phenology: Widely distributed and relatively common, the majority of specimens from southeast England, extending northwards to Yorkshire. Ireland (O'Connor *et al.*, 2007): Armagh; Mayo. Flight period: June to July.

Biology: *L. nigridens* was reported to have been reared from Psychidae (Morley: '*Psyche intermediella*' – f. Chapman). More recently, it has been bred from *Diplodoma herminata* (coll. v.1983) [NMS]. This association seems a little surprising, given the abundance with which the ichneumon occurs in some localities – plus the relatively small populations of psychid moths generally encountered in the same areas.

Lissonota palpalis Thomson, 1889

Taxonomy: *L. palpalis* is generally recognisable from the colour patterning of the head, although additional attention must be given to dimensions of the first flagellar segment.

Distribution, abundance and phenology: Widely distributed and fairly common, despite being poorly represented in older collections. The species seems most easily collected via Malaise-trapping or host-rearing. Ireland (O'Connor *et al.*, 2007): Down. Flight period: May to July.

Biology: *Lissonota palpalis* has been reared via *Ypsolopha parenthesella* (L.) and *radiatella* (*ustella* (*Clerck*)) [NMS / NHM]. Habitats are deciduous woodland and similar biotopes. There is a single rearing 'with *E. sulphurella*': Devon, Spitchwick [NHM] (see comments below).

Heterogeneity within Lissonota palpalis*:*
At an early stage, it seemed to me that *L. palpalis* falls naturally into two segregates, one having broader central tergites plus greater posterior ocellar interspace versus ocellar-ocular interspace, the other with both tergites and ocellar-ocular proportions narrower. A second avenue of division within *palpalis* sensu lato lies in the fact that host rearings also split into two categories: one attacking *Cerostoma* (Yponomeutidae) larvae feeding within leaf rolls on shrubs and trees, the other attacking oecophorid (and perhaps other 'micro') larvae living in decaying timber. With the

latter finding, it is worth noting that I have personally encountered the species inspecting dead *Pinus* timber (Norfolk, Santon Downham). A simple solution to this apparent anomaly would be that two species are involved, these being separable on both structural and biological factors. Alternatively, the 'species' in question might be two generations of a single species. Examination of a larger, more widely distributed sample of specimens led to the observation that the apparent structural differentiation between the two segregates breaks down with respect to all characters studied.

A final conclusion on the *Lissonota palpalis* question cannot be reached until much more reared material is obtained, especially with decaying timber examples – since the latter element is very poorly represented in collections at the present time.

Lissonota trochanterator Aubert, 1972

Taxonomy: The head dimensions render this species distinctive within the *buccator* group.

Distribution, abundance and phenology: Little known in the UK.

'Broc N.F.' (presumably Hampshire, New Forest, Brockenhurst), 7.viii.1909 (Morley Coll.). [NHM].

Biology: Unknown. The existing data may indicate a link to ancient deciduous forest.

Lissonota palpator Aubert, 1969 [*parasitellae*]

This species is the first in the series of the '*buccator* group sensu stricto' – a complex of closely interrelated forms that was very poorly known before the publication of Aubert's 1978 catalogue. Given that this author himself remarked that much work still needed doing, it is not surprising that much further research has had to be carried out in order to render this section workable. One contribution was made by Horstmann (2003), while the remainder forms part of the present work.

Taxonomy: *L. palpator* is at once recognisable within the *buccator* group on account of the unusually elongate central tergites of the metasoma, combined with their heavily punctate surface sculpture. The species might easily be confused with *tenerrima* (see p. 91), but can be readily separated by the different mesopleural sculpture, combined with stronger puncturation on the central abdominal tergites. In males (which superficially strongly resemble *tenerrima*), tergite 1 is very heavily punctured, whereas it is predominantly coriaceous in *tenerrima* males.

Aubert incorrectly synonymised his own *palpator* with *silvatica* of Kriechbaumer, following which Horstmann (2003) described the species as new to science.

Distribution, abundance and phenology: The species is somewhat sparsely represented in older collections – in part due to confusion with other members of the *buccator* group. In practice, *L. palpator* is frequently encountered through Malaise trapping, by which latter means it has proven fairly common in suitable localities – although perhaps being linked to older, well established habitats. Present records extend as far north as Aberdeenshire.

England: Norfolk, Santon Downham; Cheshire, Abbotts Moss; Berkshire, Burnham Beeches; Cambridgeshire, Chippenham Fen. Surrey, Ashtead Common. Scotland: Aberdeenshire, Morrone Birkwood [NMS]. Also: a few examples in NHM. Flight period: June to August.

Biology: *L. palpator* occurs amongst rotting timber, and has been reared from *Triaxomera parasitella* (Huebn.) on fungi growing on *Fagus*, *Quercus*, and *Carpinus* (f. Horstmann, 2003).

Lissonota accusator (Fabricius), 1793 [*unicincta*]

Taxonomy: The slender first, plus lightly sculptured central tergites are the important recognition characters.

Distribution, abundance and phenology: Little known in the UK.

Twenty five specimens: England: Cambridgeshire, Madingley, 11.vii.36. [NHM]. Also: Berks., Silwood Park, 1972 [UM]. Ireland (O'Connor *et al.*, 2007): Wicklow**.**

Biology: There appear to be no valid records of host relationships for this species.

Lissonota carbonaria Holmgren, 1860

Taxonomy: This and the next species are unusual in the *buccator* group on account of the weak puncturation of the central metasomal tergites. Care must therefore be taken in order to avoid confusion with the *gracilenta* group.

I have so far encountered no (authenticated) material of the male sex.

Distribution, abundance and phenology: *L. carbonaria* seems usually to be collected through host-rearing, which circumstance makes it difficult to assess distribution, abundance, or phenology. There is also a record via light-trapping. Existing records extend throughout England, northwards to Tayside in Scotland. Full data is given below:

England: two females, London, Hampstead Heath, ex *Cydia splendana* in acorn (*Quercus robur*), coll. -.x.; two females, Wiltshire, Savernake Forest, Malaise trap, 2-22.v.; Cheshire, Abbotts Moss, from indet. 'micro' larva in *Quercus* trunk trap, 11.x.; Kent, Blean, spongy *Quercus* galls, coll. 29.ix.; N. Warwicks, Bentleypark Wood, *Pinus*, 21.iv.; Devon, Bystook Woods, ex *Rhyacionia* sp., *Pinus*, 25.ii.; Kent, Church Wood, ex spongy *Quercus* galls, 29.ix.; two females 'ex *splendana* or *Pammene juliana* (*fasciana*)', Cambridgeshire, Chippenham Fen; Merseyside, Wirral, Hilbre Island, at red light [NHM]. Scotland: E. Lothian, Whitadder Reservoir, 2.v. [NMS]. 'ex *Retinia*'; Ireland (O'Connor *et al.*, 2007): Dublin; Wicklow – requiring confirmation in view of recent taxonomic revisionary work. Flight period: May and June.

Biology: On the basis of verified data detailed above, the species attacks *Cydia splendana* (Huebn.) in *Quercus robur* acorns, as well as 'spongy oak galls' – also *Rhyacionia* on conifers. In addition, *Retinia resinella* (L.) may be a further target species. Habitats clearly include both deciduous and coniferous woodland.

The male sex is so far unknown to the author.

Species allied to carbonaria:
The next two species are closely related to *L. carbonaria* – and as already indicated, it is useful to have adequate comparative material available in order to build confidence in species determination. This situation is further complicated by the fact that all three are rarely collected

other than via rearing from hosts. This could, however, be at least partly due to the fact that an early flight period is involved – when collecting activity tends to be at a lower profile.

Lissonota simulator sp. nov.

Female: Flagellum 1 from 4.5-5 times longer than broad. Temples distinctly narrowing in dorsal aspect – although noticeably convex, lateral pubescence long and dense. Interocellar distance 1.5-1.7 x posterior ocellus to eye and 0.6-0.75 x hind ocelli to occipital carina. Maximum temple length a little less than to approximately same length as flagellum 1. Minimum genal length 1.5-1.7 x malar space. Malar space 0.8, to approximately equal to basal width of mandible. Thorax: fore wing length 4-6 mm. Mesopleurum distinctly coriaceous between punctures. Speculum small (extending at most about 0.35 distance to epicnemium), closed by 3-4 rows of punctures. Propodeum: posterior carina weak to normal. Metasoma: Tergite 1 from 1.25-1.4 x longer than broad, sculpture: transversely coriaceous, almost impunctate, pre-apical impression weakly defined, striation confined to lateral aspect. Tergite 2 transversely coriaceous, almost impunctate, slightly more than, to about 1.3 x broader than long. Tergite 5 similarly sculptured to tergite 4, tergite 6 uniformly pubescent. Ovipositor: a little shorter, to a little longer than metasoma. Colour: black: thorax at most with small, pale 'shoulder marks'; legs testaceous, hind tarsi more or less darkened.

Male: From female: mesopleurum weakly coriaceous, speculum larger, not 'closed'.

HOLOTYPE. ENGLAND: female, Devon, Zeal Monachorum, ex *Aloboria geoffrella* (L.) on *Rubus fructicosus*, coll. 27.iii.2010, emerged by 29.iii.10 [NMS].

PARATYPES. ENGLAND: three females, Worcestershire, Solihull, ex *Esperia sulphurella* (F.) in dead *Prunus*, coll. 17.iv.1976, em: 25, 28, iv. and 6.v.76; one female: Yorkshire, Sheffield, ex *E. sulphurella*, em. iii.1989; one male: Dover, Shakespeare Cliff, ex *E. sulphurella* in dead wood, 18.iv.1987; one male: Hampshire, Botley Wood, ex *Aloboria geoffrella* (L.) in dead *Rubus fructicosus*, coll. 3.iii.2002, em. iv.02; Kent, E. Blean, Childs Forest, ex dead stems of *Hedera*, coll. 11.iv.1991 [NMS]. One female, Surrey, Wimbledon Common, ex *E. sulphurella*, iv.1947 [NHM]. **SCOTLAND:** one male, Midlothian, Pentland Hills, ex larva in wooden post, ix.1976 [NMS].

Taxonomy: *L. simulator* will run to *carbonaria* in earlier keys, but is distinguished on the characters given here.

Distribution, abundance and phenology: Known almost exclusively from the type specimens. Flight period: the only wild-caught example was found during early May (see above).

Biology: While it has to be said that *Esperia sulphurella* is commonly cited as being the 'suspect host' for any *Lissonota* collected around rotten timber, sources for the present determination can probably be taken as being reliable. Additional rearings are from *Aloboria*, perhaps implying a general association with oecophorid larvae living in rotten wood. There thus appears to be a biological differentiation between this and the previous species – *carbonaria* attacking hosts in fruits and gallular growths, *simulator* being linked to rotten timber. The species has been taken at light.

Lissonota arborator sp. nov.

Female: Flagellum with segment 1 from 4.5 to over 6 x longer than broad, segment 2 about 3.5 x. Temples in dorsal aspect with contour narrowing, lateral pubescence fine. Interocellar distance about 1.65 to around 2 x posterior ocellus to eye, and approaching 0.75 x distance between posterior ocellus and occipital carina. Vertex and frons with obscure puncturation. Maximum temple length about same length as flagellum 2. Minimum genal length much greater than malar space, somewhat greater than 0.33 x eye length: malar space about 1.5 x basal width of mandible. Fore wing length 3.5-4.5 mm. Nervellus subvertical. Mesopleurum punctate, the interspaces coriaceous; speculum with impunctate zone extending about 0.3 x distance to epicnemium. Propodeum almost impunctate, with lateral and posterior carinae. Metasoma: tergite 1 about 1.5 x longer than broad – transversely coriaceous, with some indistinct puncturation. Tergite 2 punctate, the punctures mostly approximately equal to interspaces. Tergites 2 and 3 distinctly broader than long. Tergite 5 sculpture similar to that on tergite 4; tergite 6 with pubescence continuing across dorsum. Ovipositor from somewhat shorter, to noticeably longer, than wing length. Colour: black: head with yellow spot on vertex, *plus a very short (almost quadrate) yellow line annexed to antennal scrobes*. Thorax at most with small, broadly triangular yellow shoulder marks, scutellum with small yellow lateral mark. Abdomen piceous, central tergites with paler apical margins. Legs testaceous, tarsi darker.

HOLOTYPE. ENGLAND: female, Devon. Canonteign, ex *Schiffermuelleria grandis* (Desv.), under *Hedera* bark, 4.iv.1998. [NMS].

PARATYPES. ENGLAND: four females, Hampshire, Romsey, Aubridge, dead *Larix* trunk (at 6, 8 and 9 m above ground level), 25-30.iv.1984. [NHM]. One female, Berkshire, Silwood Park, ex *Esperia sulphurella* (F.), dead *Salix* (as cocoon), 24.iv.2001. One female: Hampshire, Havant Thicket, ex dead *Quercus*, 2.iii.1996]. Two females, London, Richmond Park, canopy fogging (*Quercus*) 15.v.1984 [NMS]. One female, London, Sydenham Wood, Malaise trap, 11-24.v.1993 [HM].

Taxonomy: The yellow vertical and frontal marks are distinctive; however, in common with other species, these pale markings are sometimes absent, thus structural features have to be taken into consideration for accuracy in identification.

Distribution, abundance and phenology: Probably a species that is most likely to be encountered through host-rearing. Flight period: the only wild-caught examples were taken during May.

Biology: While an association with *Schiffermuelleria* is evident, the rarity of this host clearly indicates that a wider range of hosts must be targeted (probably including *Esperia*). It is perhaps notable that the species has been collected in urban areas, in one instance via canopy-fogging, and in another from log samples taken at considerable heights above ground level.

Lissonota folii Thomson, 1877 [*transversa*]

Taxonomy: Typical examples with richly yellow-marked face and thorax are not at all difficult to distinguish from other members of the subgroup (provided that structural traits are also taken into account). In addition, tergites two and three generally have a testaceous apical margin, with the latter frequently widening laterally, and sometimes with the central tergites quite widely suffused with testaceous. However, *folii* is a very variable species, and the best advice here is to accumulate a reasonably sized collection of *buccator* group material prior to any attempt at building confidence in dealing with infra-specific variation of this magnitude.

Males have a yellow-marked face, vertex, pronotal collar, shoulder marks, front and middle coxae and trochanters – the mesopleuron sometimes also with extensive yellow area.

Distribution, abundance and phenology: Very common and widely distributed, with records extending to Easter Ross. Ireland (O'Connor *et al.*, 2007): Donegal. The species occurs also in the Nearctic region, extending northwards into Canada (Townes and Townes, 1978). Flight period: June to August.

Biology: *L. folii* is a species associated with foliage of deciduous trees, occurring in any suitable locality – including gardens, and sometimes attracted in numbers to honeydew. Despite its ubiquitous distribution and general abundance, rearing records are extremely sparse. It has been reliably recorded from *Epinotia brunnichana* (L.), also from *E. sordidana* (Huebn.) [NMS]. Additionally, one specimen has been reared from *Zeiraphera griseana* (Huebn.) [NHM]. The Nearctic tortricid *Z. fortunana* is given as a host by Townes and Townes (1978). Males have been collected in swarms, usually around tree trunks.

Lissonota serena sp. nov.

Female: Flagellum 1 with length:breadth ratio 4.5-5. Temples in dorsal aspect with contour narrowing – although more or less convex. Occipital carina weakly angled. Sculpture of vertex: shining, very weakly coriaceous with distinct punctures, the latter at most equal to interspaces. Frons: as vertex, but more heavily coriaceous and somewhat dull. Interocellar distance approximately 1.3 x posterior ocellus to eye, and 0.66 to 0.75 x posterior ocellus to occipital carina. Maximum temple length approximately equal to length of flagellum 1. Minimum genal length 0.66 to 0.75 x malar space; malar space 1.3-1.5 x basal width of mandible. Fore wing length 6-7 mm. Mesopleurum: speculum very small, reaching less than 0.25 length of pleurum, more or less obscured by puncturation. Mesosternal sulcus with the costae very strong, ridge-like in posterior half. Propodeum: coriaceous, uniformly punctate, posterior carina strongly developed. Metasoma: tergite 1 about 1.65 x longer than broad, coriaceous with distinct punctures, latter mostly narrower than interspaces – the punctures absent along dorsal longitudinal depression. Tergite 2 subquadrate, with punctures predominantly narrower than interspaces. Ovipositor: from somewhat longer than metasoma plus propodeum, to greater than body length. Colour: head, thorax and metasoma black, with no pale markings; legs red, with all coxae and trochanters blackish.

Male: From female: tergite 2 with punctures approximately same diameter as interspaces in front portion.

HOLOTYPE. ENGLAND: female, Berkshire, Windsor Forest, 24.vii.-23.vii. [HM].

PARATYPES. ENGLAND: three females, Surrey, Headley Warren, Malaise trap, 23.vi.-14.vii.2000; Buckinghamshire, Windsor Forest, 24.vii.-23.viii.1997 [HM]; female, Oxfordshire, Dry Sandford Pits, Malaise trap, 30.vi.1990; **SCOTLAND:** female, Edinburgh, Blackford, ex *Dichrorampha*, em. 4.viii.1992 [NMS].

In addition to type material, there is a 'probable' male: Buckinghamshire., Burnham Beeches, vii.05, Rothamsted M.V. trap [NHM].

Taxonomy: Leg colour forms an 'easy' character for recognition of this species. However, this is an unsafe trait to rely upon, given the usual range of variation in pigmentation in banchines (as indeed, in most other Ichneumonidae). Bearing in mind the fact that only a very few specimens have so far been collected, it is impossible to fully assess the reliability of this particular trait.

Distribution, abundance and phenology: Known only from the above listed material. This could well be a species that is widespread, but rarely collected other than by Malaise-trapping or host-rearing.

Biology: *Lissonota serena* has been reared from *Dichrorampha*, feeding at the roots of Ox-eye Daisy (*Leucanthemum vulgare*). This indicates a link to open country, such as commons and similar habitats – which is in stark contrast to other *buccator*-related species. Several members of the host genus are widespread and common (albeit of cryptic habit in the larval stage).

Species allied to buccator *Thunberg:*

As with the *carbonaria* complex, it is useful to have good comparative material at hand in order to gain confidence in species recognition. In this particular instance, specimens are relatively easy to collect by searching around rotten timber. The application of a Malaise trap is also an advantage. The flight period lies more towards summer, thus the insects are much less likely to be overlooked, than are those species related to *carbonaria*. So far as the male sex is concerned, I have encountered insufficient (confirmed) material, therefore it has not been possible to give much guidance in separating species. The name '*errabunda*' is an '*in lit.*' sobriquet for this species complex.

Lissonota punctiventrator Aubert, 1977

Taxonomy: The short first flagellar segment renders *L. punctiventrator* distinct from the succeeding species. (I have seen a few specimens with this segment somewhat more elongate than the dimensions given in the key).

Distribution, abundance and phenology: Relatively uncommon, although occurring at least as far north as Tayside.

England: 'Ryde' (Morley Coll.), Desvignes, Stephens [all NHM, det. Aubert]. Malaise trap material from: Oxfordshire; Wiltshire, Savernake; Norfolk; Cambridgeshire; Berkshire. Wales: Wrexham, Glyn Ceiriog. Scotland: Stirlingshire, Flanders Moss; Dunbartonshire, Caldarvan; Tayside, Perthshire, Errol, Paddockmuir Wood. [NMS]. Flight period: July and August, into September.

Biology: Hosts unknown, perhaps associated with rotten timber. *L. punctiventrator* has been taken in woodland habitats in general – possibly with two generations per annum, on the basis of existing data.

Lissonota ?buccator (Thunberg, 1822)

Taxonomy: Typical examples of *buccator* are easily recognised within the species complex on account of the distinctive facial markings. More melanistic individuals need to be examined carefully in order to avoid confusion with others (and especially with *L. punctiventris* – see p. 107). In the latter case, the broad central metasomal tergites offer a good recognition characteristic. The few male specimens encountered agree with the female sex with respect to the length:width ratio of the central metasomal tergites.

It should be noted that interpretation of *Lissonota buccator* Thunberg has had to be based on Horstmann (2003), since the type itself is unobtainable at present. Unfortunately, *L. buccator* sensu Horstmann is split into two segregates in the present study, one of which is here interpreted as part of *punctiventris*. Since Horstmann did not indicate whether the characteristic facial colour pattern of *buccator* (as now recognised) appertains to the type specimen, it is not possible to make a final judgement on the application of the Thunberg name.

Distribution, abundance and phenology: Widely distributed, probably common locally. Ireland (O'Connor *et al.*, 2007): Dublin; Wicklow; the Irish records refer to a 'composite' species – thus require confirmation. Flight period: June to August.

Biology: Examples have been reared in association with fungi growing on rotten timber (*Calvatia, Pleurotus*). In a few instances, reared hosts have been given as *Nemopogon* and *Triaxomera* (Tineidae). The only other apparently authentic rearing record is from *Ypsolophus parenthesella* on *Quercus*: Norfolk, Diss, host collected 25.ix.1994 (see also comments on *L. palpalis* regarding the anomaly of a single parasitoid attacking widely different hosts feeding on a different substrate – also *L. biguttata*) [NMS].

Lissonota punctiventris Thomson, 1877

Taxonomy: Horstmann (2003) discusses discrimination of *punctiventris* from *L. buccator*. In the present study, *buccator* sensu Horstmann is made-up of *buccator* itself plus a component here recognised as *punctiventris partim*. The dimensions of the central tergites form the recognition trait for *buccator* as interpreted here (see above), while the two characters used by Horstmann in defining his *punctiventris* (size of malar space versus mandible base, plus presence / absence of impunctate zones on tergite 2) can be found in all possible combinations within contiguous populations. Resolution of this problem has been based on long series of specimens collected from several localities.

Only inadequate material of the male sex has so far been examined. In these, the face, pronotum, fore, and mid- coxae are yellow. Due to lack of male-female associated rearings for both *buccator* and *punctiventris,* it is not possible to separate males of the two species with a satisfactory degree of confidence.

Presence of a more or less striate impression over tergite 2 may cause confusion with the saturator group (see remarks under latter heading).

Distribution, abundance and phenology: *L. punctiventris* is widespread, and can be common in suitable habitats. Older records are sparse.

England: Ryde (Morley); one Desvignes, one Stephens [NHM]. Devon, Dewerstone, ex dead ?*Quercus* bark with *Triaxomera fulvimetrella* (Sodof.). Scotland: May/June, Tayside, Paddockmuir Wood, Malaise trap. Also: 'observed ovipositing on *Pleurota*' [NMS]. Flight period: May to June and August to September (probably two generations a year).

Biology: *L. punctiventris* is associated with rotten timber, and is seldom collected any significant distance from this pabulum. The species has been reared from *Triaxomera*, repeatedly from *Nemopogon cloacella* (also *N. personella*) [NMS]. Morley (1908) gives '*Orchesia micans*', no doubt in error. The usual biotope is deciduous woodland.

Genus *Cryptopimpla* Taschenberg, 1863

The present genus might arguably be placed as a subgenus of *Lissonota*. Townes and Townes (1978) placed *Cryptopimpla anomala* in the latter genus, since the species is said to agree with *Lissonota* in having the speculum mostly impunctate and the female flagellum without moniliform distal segments. However, some 'true' *Lissonota* have a punctate speculum, and *Cryptopimpla anomala* does in fact have a few flagellar segments weakly barrel-shaped. A better case might be made for exclusion of *C. arvicola* from the genus.

Cryptopimpla species generally occur in open country, including northern moorland and mountains – also chalk downs in the south. None can be said to be generally distributed and common. For the most part, they prefer a 'quality' environment, as distinct from secondary habitats or disturbed ground, such as are favoured by many of the commoner banchine species. Unlike *Lissonota*, their host preferences generally lie with exophagous larvae (Geometridae).

Males are difficult to place as *Cryptopimpla* – for which reason it is best to limit reference to the female sex in the present treatment (this state of affairs is by no means unique within Ichneumonidae – cf. for example, the campoplegine genera *Diadegma* and *Hyposoter*, which are distinguished purely on ovipositor length).

Size: fore wing lengths range from 4-6 mm, usually towards the lower end of the stated range.

Key to species of *Cryptopimpla*

1. Speculum present, large and entirely glabrous (Fig. 272); mandible base greater than 2 x width of malar space, genal carina almost reaching mandible base (Fig. 273); female flagellum with last 12 or so flagellar segments of usual cylindrical form – antenna gradually tapering towards apex, setae longer, tyloids prominent (their colour contrasting with background) (Fig. 274); propodeum with posterior carina weakly indicated, metasoma predominantly black. *East Anglia, rare* .... ***arvicola*** (Grav.) [*brachycentra*] (p.110)

– Speculum absent, or much reduced (Fig. 275) – or else with conspicuous microsculpture in anterior region; mandible base narrower than 2 x width of malar space, genal carina often meeting hypostomal well behind mandible base (Fig. 276); female distal flagellar segments more or less moniliform (either barrel-shaped, spindle-shaped, or of hexagonal contour) (e.g. Figs 278, 279) – flagellum thus non-tapering, its setae of normal length, tyloids less prominent and tending to be same colour as main body of segment; if propodeum lacking posterior carina, metasoma predominantly red 2

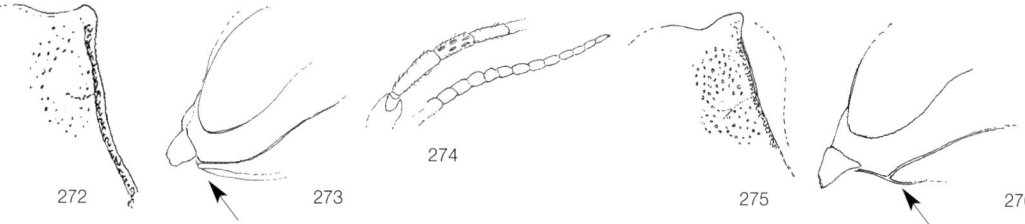

272   273   274   275   276

2. Metasoma predominantly black; female distal flagellar segments spindle-shaped, and with at least a weakly defined central ridge 3

– Metasoma predominantly red; if female distal flagellar segments spindle-shaped, then with no trace of central ridge (Fig. 277) 4

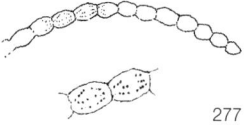

277

3. Distal flagellar segments of female with strongly-defined central ridge, thus hexagonal in profile (Fig. 278); minimum genal length not greater than 1.5 x malar space; all trochanters / trochantelli blackish, fore and mid coxae dark suffused in female, black in male (which latter have face, mandibles, notum and scutellum lacking yellow markings). *Rare* .... *caligata* (Grav.) (p. 110)

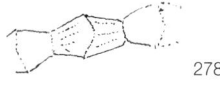

— Distal flagellar segments with only a weakly-defined central ridge (Fig. 279); minimum genal length about 2 x width of malar space; trochanters / trochantelli reddish, fore and mid coxae red, yellow-marked in male (latter with face, mandible, notum and scutellum also yellow-marked). *Sometimes common* .... *calceolata* (Grav.) (p. 111)

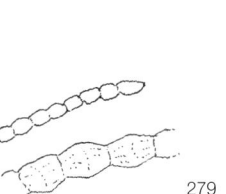

4. Propodeum with neither lateral nor posterior carinae (males with trace of laterals); tibial spinules strong (Fig. 280); mesopleurum glabrous to weakly coriaceous between punctures, latter predominantly narrower than interspaces; tegulae black, all coxae, trochanters, hind femur, black – hind tibiae and tarsi dark; (distal female flagellar segments moniliform (see Fig. 277)). *Widespread* .... *errabunda* (Grav.) (p. 111)

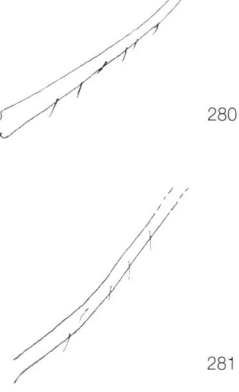

— Propodeum with both lateral and posterior carinae; tibial spinules weak (Fig. 281) legs extensively reddish; (mesopleurum often distinctly coriaceous between punctures – latter mostly not less than diameter of interspaces, tegulae sometimes pale) 5

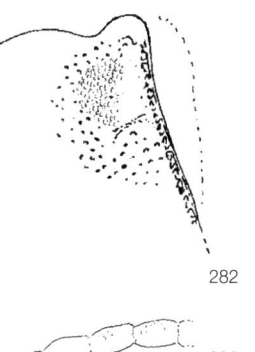

5. Speculum large, with a region of coriaceous microsculpture in front (Fig. 282) ; flagellum with some distal segments weakly barrel-shaped (Fig. 283); mandible base approximately equal to malar space; mesopleurum coriaceous between the punctures; (males with long pubescence on head). *Northern* .... *anomala* (Holmgren) (p. 111)

— Speculum small or absent; flagellum with many distal segments strongly barrel- or spindle-shaped (Fig. 284); mandible base usually distinctly unequal to malar space; (mesopleurum sometimes lacking microsculpture between punctures; males with normal cephalic pubescence) 6

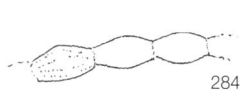

6. Tergite 1 predominantly black; mesopleurum more or less glabrous between punctures; malar space greater than width of mandible base; ovipositor longer than basitarsus 3; flagellum 2 about 2.5 x longer than broad. Males with face yellow along eye margins alone. *Northern* .... *hertrichi* Heinrich (p.112)

— Tergite 1 entirely red in female sex; mesopleurum coriaceous between punctures; malar space narrower than width of mandible base; (ovipositor shorter than hind basitarsus; flagellum 2 up to about 3 or more x longer than broad; known males with face more extensively yellow-patterned) 7

7. Claws obtusely curved and slender (Fig. 285); genal carina meeting hypostomal far behind mandible base (up to 0.4 x width of latter) (cf. Fig. 286); interocellar distance at least equal to ocellus to eye. *Southern, rare* .... *quadrilineata* (Grav.) [*blanda*] (p. 112)

– Claws almost rectangularly bent and more robust (Fig. 287) ; genal carina inflexed at most 0.33 x mandible width; interocellar distance at most equal to ocellus to eye distance. *Apparently commoner in north* .... *altipes* (Holmgren) (p. 112)

285

286

287

Species accounts for *Cryptopimpla*

Cryptopimpla arvicola (Gravenhorst, 1829) [*brachycentra*]

Taxonomy: As already mentioned above, this species might well be included in *Lissonota*, due to the unmodified distal flagellar segments. Males have a yellow orbital line. The placement here is partly a matter of convenience (although host-selection factors might support a genuine relationship with *Cryptopimpla*).

Distribution, abundance and phenology: Rare, a few records from East Anglia and Essex.

England: Norfolk, Coston, ex *Ligdia adusta* on *Euonymus* (Haggett) [NMS]; 'Brit Is.' (Capron); W. Suffolk, Tuddenham, 1914; three specimens, Essex, Colchester (Harwood) [NHM]. Flight period: no wild-caught adults carrying relevant data have been encountered.

Biology: Information on the species is somewhat fragmentary, apart from the rearing data given above. The Bridgman Collection [NCM] contains material reputedly reared via *Anticlea badiata*. It seems likely that *C. arvicola* is to found in relatively open country.

Cryptopimpla caligata (Gravenhorst, 1829)

Taxonomy: The two black-bodied species of *Cryptopimpla* are not difficult to distinguish, following characters given in the key.

Distribution, abundance and phenology: *L. caligata* is poorly represented in collections. A few 'historical' specimens (lacking provenance); England: Yorkshire (det. Perkins) from: Frog Hall, 1930; Hovingham, 1935; Crimsworth Dene, 1935 [NHM]. Ireland (O'Connor *et al.*, 2007): early 20th century – Down; Wicklow. Recent record: Yorkshire, Grimston Moor, 2011 [WE]. Flight period: late July, into September.

Biology: Reputed host: *Anticlea badiata* (D. & S.), from Morley (1908, citing Bridgman / Bignell). This is another species for which a limited amount of collecting data can lead to little more than speculation as to species habitat.

Cryptopimpla calceolata (Gravenhorst, 1829)

Taxonomy: See comments on previous species.

Distribution, abundance and phenology: Little known in the past, although recently taken relatively frequently in parts of England (distribution extends to Scotland).

England: a number of 'historical' specimens, all lacking provenance. More recently: Berkshire, Silwood [NHM]; nine males, four females, Surrey, Banstead Down; Kent, Folkestone Warren, Malaise trap; Chobham; Headley Warren [HM]; Yorkshire, Thorpe Marsh [WE]; Sussex, Maids Common (Haggett); Norfolk, Santon Downham. Scotland: Fife, Dumbarnie Links, dunes [NMS]. Flight period: an autumnal species; September, through to early October.

Biology: There appears to be a habitat link with chalk downs and heaths, also dunes. The recorded host is *Camptogramma bilineata* (L.) (Haggett – see above). [NMS]

Cryptopimpla errabunda (Gravenhorst, 1829)

Taxonomy: The morphology of the propodeum renders this species easily recognisable. However, in males, the carina lateralis may be more or less in evidence.

Distribution, abundance and phenology: Widespread in England, although relatively uncommon in occurrence.

England: older material – Devon, Cornwall; West Suffolk, one reared from *A. sinuata* [NHM]; also: Norfolk, Thomson Common, *Eppirrhoe alternata* (Muell.), coll. 10.ix.; Cambridgeshire, Hampshire, [NMS]. Berkshire, Surrey [HM]; Yorkshire [WE]. Ireland (O'Connor *et al.*, 2007): widespread. Flight period: late June, into August.

Biology: *C. errabunda* has been reared from: *Eppirhoe alternata* (see above). Amongst other reputed rearings is one from Brischke via *Cidaria sp.* (see Morley, 1908).

Cryptopimpla anomala (Holmgren, 1860)

Taxonomy: The unusually large mesopleural speculum plus weakly moniliform female flagellar segments form good recognition characters for *C. anomala*.

Distribution, abundance and phenology: Widely distributed in upland areas in Scotland – including Aberdeenshire, Inverness-shire, Perthshire, Wester Ross, Jura, Sutherland, and Kirkcudbrightshire. The habitat extends to high altitude (certainly up to 760 m), and the species has been found immobilised on snow. England: Moor House Nature Reserve; Yorkshire [WE]; Pen-y-Ghent, about 2000 ft [NHM]. Wales: Radnor Forest, above 1500 ft. Scotland: reared 'ex *caesiata*' (D. & S.), Braemar. [NHM]. Wester Ross, An Teallach, 530 m, damp *Calluna* heath, Malaise trap; one male, Kirkcudbright, Southerness, ex *Ematurga atomaria*, coll. 11.vii.; one female, Abernethy Forest, ex *Eulithis testata* (L.), coll. 17.v. [NMS]. Flight period: on the wing during April, through to July.

Biology: The hosts cited above are often common on upland heath and moorland. Collecting data are associated with native Pine, also *Betula*, *Ulex* – the hosts' larval stages occurring on *Calluna, Erica* and *Salix*.

Cryptopimpla hertrichi Heinrich, 1952

Taxonomy: The mesopleural sculpture is distinctive within this section of the genus.

Distribution, abundance and phenology: An upland and montane species, widespread in Scotland, also occurring in suitable localities in northern England.

England: one male, one female, Durham, Weskerley, ex *Entephria caesiata* (D. & S.). Scotland: Angus, Caenlochan Crags, Glas Maol, 2500 ft; Aberdeenshire, Glen Muick, on *Betula nana*; one male, one female, Dunbartonshire, Lang Craigs, ex *Entephria flavicincta* (Guen.); one male, Perthshire, Ben Lawyers NNR, ex *Entephria flavicincta*; two males, one female, Orkney, Orphir, ex *Entephria caesiata*; Aberdeenshire, Morrone Birkwood NNR, ex *E. flavicincta*; nine females, Speyside, Abernethy Forest, Malaise trap, vii.-x. [NMS]. Flight period: July through to October, reared through hosts collected between March and June.

Biology: *C. hertrichi* is clearly another montane insect, apparently less common than the previous species. The host genus *Entephria* is associated with herbaceous plants growing in upland localities in the more northern parts of the UK.

Cryptopimpla quadrilineata (Gravenhorst, 1829) [*blanda*]

Taxonomy: This and the next species are closely allied, and care is needed in order to ensure that accurate observations are made with characters given in the key.

Distribution, abundance and phenology: A Holarctic species, *C. quadrilineata* is treated in the context of three subspecies in the Nearctic region (Townes and Townes, 1978). It is of rare occurrence and southern distribution in the UK.

England: a few 'historical' specimens – 'Eaton, Morley'; also: 'Wales: Colwyn Bay' [NHM]. Recent data: England: Cornwall, Kennack Sands, ex indet. geometrid on *Thymus*, coll. 24.iv.1992; Oxfordshire, Frilford Heath, 18.vi-12.vii.1991, Malaise trap. Flight period: late July, into August.

Biology: *C. quadrilineata* has been reared from an unknown geometrid species. It is difficult to draw any conclusions on habitat type on the basis of the very limited information available.

Cryptopimpla altipes (Holmgren, 1860) Plate 14

Taxonomy: See comments on previous species.

Distribution, abundance and phenology: Apparently commoner in the north – although found also in East Anglian fens and in Oxfordshire. There are a few records amongst 'historical' material.

England: some 'historical' material from Desvignes / Stephens (no provenance) [NHM]. More recent captures (all via Malaise trap): England: Cambridgeshire, Chippenham Fen; Oxfordshire, Wychwood Forest. Norfolk, Santon Downham, heath with *Betulus* and *Pinus*. Scotland: Aberdeenshire, Braemar, Morrone Birkwood, *Betula, Juniperus, Populus tremula* wood; Inverness-shire, Strathfarrar, native *Pinus*; Loch Arkaig, native *Pinus*; W. Ross, Beinn Eighe NNR, native *Pinus*; Rassal NNR [NMS]. Flight period: May through to August (suggesting two generations per year).

Biology: Host unknown. As with the two preceding species, there is a definite habitat link with upland moors and forest in Scotland – albeit with surprising additional records from East Anglia (fens) and Oxfordshire (woodland).

*ADDENDUM – **Cryptopimpla** sp. indet:*

The following account is a reference to an apparent *Cryptopimpla* species which has so far eluded identification, owing to the fact that the female sex has not as yet been encountered.

The present *Cryptopimpla* species is easily distinguished from other black-bodied kinds on account of having unusually broad basal metasomal tergites – tergite 1 being only about 1.6 x longer than broad. Tergite 2 is strongly transverse: about 1.4 x broader than long (first tergite at least 2 x, second at most a little broader than long, in other black-bodied *Cryptopimpla* species). A further unique trait lies in the complete lack of the usual series of transverse costae running along the mesosulcus. In addition, the species is distinctive on the basis of its small size (fore wing length only about 4.5 mm, as against the usual range of 5-6 mm in *Cryptopimpla*). The mesopleural speculum is large, although at the same time bearing a finely coriaceous sculpturing, much as in *C. anomala* (absent in other *Cryptopimpla* species – apart from *arvicola*, in which the speculum is entirely glabrous). Lastly, the posterior transverse carina of the propodeum is more or less absent (thus approaching the situation found in *C. errabunda*). So far as colour pattern is concerned, the trochantelli and hind tibial base are yellow (not so in other members of the 'black group').

The hosts of species indet. are Geometridae feeding on flowers of Rosaceae: England: Norfolk, Foulden Common, from *Chloroclystis rectangulata*. Cornwall, Bodelva, Par, from *C. rectangulata* on *Malus*. Lancashire, Gait Burrows, from *C. rectangulata* or *chloerata* on *Prunus*. Berkshire, Silwood Park, from *C. chloerata* on *Prunus*. Host larvae were collected during May in southern England, June in the north.

While the above information readily distinguishes the present from other *Cryptopimpla* species with black metasoma, it will nevertheless be necessary to locate female examples before a formal description can be made (and indeed, to confirm generic placement itself).

Tribe Banchini

The tribe Banchini had been grouped with the 'Ophioninae' according to the classical authors. Morley (1908) proposed moving *Exetastes* and *Banchus* to Pimplinae (sensu lato), as well as placing them adjacent to *Lissonota* and *Glypta*. Although Townes (1969) continued to place the two genera in the same tribe, they are in fact, quite distinct *inter se*, such that their discrimination needs no extended discussion. It should also be said that, apart from their shared anatomical similarities, the biology of *Banchus* and *Exetastes* is fairly uniform. Both attack exophagous Lepidoptera, within the superfamily Noctuoidea. In addition, they share the usual banchine trait of oviposition into early instar host larvae – and agree with one another in the habit of killing the host after a subterranean chamber has been constructed (Townes and Townes, 1978). Wahl (1988) discusses relationships of Banchini on the basis of larval characteristics.

Key to the genera of Banchini

1. Mandible with the teeth approximately of same width and both of normal build; prepectus present, scutellum with no spine (Fig. 288); second metasomal tergite with laterotergite not less than 6 x longer than broad, completely defined from tergite and folded horizontally; claws not pectinate; temple widest at vertex; (fore wing lengths 7-10 mm) .... ***Exetastes*** (p. 114)

—   Mandible with upper tooth very much wider than lower, and with indication of subdivision near top (Fig. 289); prepectus absent, scutellum usually with an apical spine (Fig. 290); second metasomal tergite with laterotergite at most a little over 2 x longer than broad, often defined from tergite by an incomplete crease, hanging almost vertically; claws pectinate; temple widening towards lateral region; (fore wing lengths usually around 7-12 mm) 2

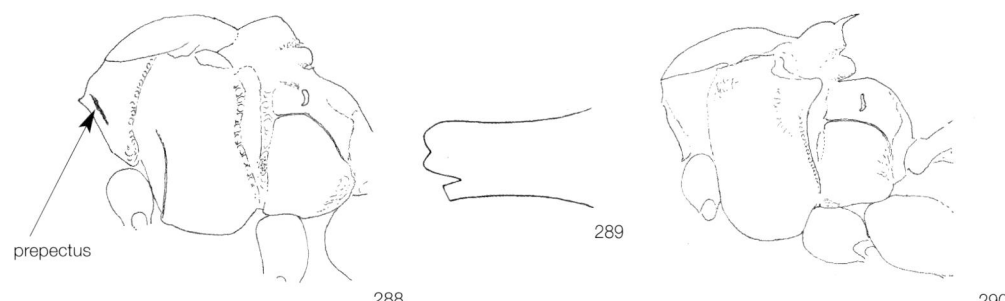

prepectus

288

289

290

2. Propodeum with posterior transverse carina present (Fig. 291); metasoma more or less laterally compressed, at least in females; (scutellum usually with a spine) .... ***Banchus*** (p. 121)

—   Propodeum with no trace of posterior transverse carina; metasoma not compressed; (scutellum with no spine) ..... ***Rynchobanchus*** (p. 126)

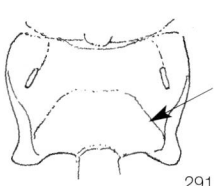

291

Genus *Exetastes* Gravenhorst, 1829

This is a large genus, especially well represented in Holarctic and Nearctic regions, with around 50 in the latter (Townes and Townes, 1978). *Exetastes* has 38 Western Palaearctic species (f. Aubert, 1978); 21 are covered by Kuslitsky (1981) for the European part of the former USSR. Hosts are usually smooth-skinned noctuid larvae which pupate in soil, apparently only one generation per annum. Nearctic *Exetastes* species occur on open country, including semi-desert and almost bare ground (Townes and Townes, 1978). In Britain, a few occasionally appear in urban gardens.

The Townes study recognises several species groups, which are admittedly artificial (and also somewhat 'weak' in terms of applicability to the UK fauna). The present treatment is mostly 'eclectic'.

There appears to have been some decline in abundance in the UK with certain *Exetastes* species. At the same time, the appearance / reappearance of one or two species in urban gardens is quite mystifying.

Key to species of *Exetastes*

1. Metasoma entirely black; interocellar distance always less than hind ocellus to eye 2

– Metasoma at least centrally red (interocellar distance often greater than or equal to distance between hind ocellus and eye) 6

2. Antennae shorter (less than 60 segments); ovipositor at most 0.6 x basitarsus 3; propodeum often lacking distinct lateral carina (tibia 3 at least partly red and / or flagellum/tarsus 3 with white ring). *Summer flight period* 3

– Antennae longer (more than 65 segments); ovipositor at least 0.6 x basitarsus 3; propodeum with lateral carinae at least partly defined (Fig. 292); tibia 3 all black; flagellum/tarsus 3 with no white ring. *Autumnal* 5

292

3. Flagellum and tarsus 3 lacking white ring; flagellum 2 at most 2 x longer than broad; interocellar distance much greater than 0.5 x ocellus to occipital carina; minimum genal length at least equal to flagellum 2, temples swollen behind (Fig. 293); tergite 2 with distinct puncturation laterally (Fig. 294); propodeum with convex dorso-median area, which may be bounded by longitudinal carinae (Fig. 295); (female: metasoma not compressed – tergite 2 transverse to a little longer than broad; propodeum with trace of lateral carinae). *Rare* .... ***fornicato****r* (F.) (p. 117)

293

294

– Flagellum and tarsus 3 with white ring; flagellum 2 often greater than 2.0 x longer than broad; interocellar distance less than 0.5 x ocellus to occipital carina; minimum genal length at most equal to flagellum 2; temples often narrowing behind; tergite 2 with at most a few fine punctures; propodeum evenly convex on dorsum; (female: metasoma often compressed) 4

295

4. Hind tibia at least partly red; area superomedia at most vaguely indicated; metasoma with tergites 5 and 6 strongly laterally compressed; tergite 2 quadrate to longer than broad, tergite 4 longer than broad (Fig. 296); mesonotum glabrous between punctures; front margin of clypeus at least same length as side margins (Fig. 297); ovipositor over 0.4 x hind basitarsus. *Formerly abundant, now rare* .... ***atrator*** (Förster) [*cinctipes*] (p. 118)

296

– Legs entirely black (excluding white hind tarsal ring); area superomedia partially defined by distinct, raised carinae; metasoma only weakly laterally compressed; tergites 2 to 6 broader than long; mesonotum dull and coriaceous between punctures; front margin of clypeus much narrower than lateral (Fig. 298); ovipositor shorter than 0.4 x hind basitarsus. *Rare* .... ***Ilyricus*** Strobl (p. 118)

297

298

5. Minimum genal length nearly 0.66 x maximum temple length; propodeum with dorsal carinae indicated; femora black [male: coxae black with yellow markings, face extensively yellow-patterned]. *Rare* .... *maurus* Desvignes (p. 118)

–   Minimum genal length much less than 0.66 x maximum temple length; propodeum with dorsal carinae tending to be confused by rugosity; femora red [male: fore coxae red, more or less yellow-marked, face marked laterally with yellow]. *Rare* .... *calobatus* Grav. (p. 119)

6. Face entirely, and mesonotal shoulder marks yellow; coxae black, yellow-marked ventrally; (interocellar distance less than ocellus to eye; tergite 1 at least 2.5 x longer than broad; clypeus and mandibles yellow) .... *atrator* (F.) [cinctipes] male (p. 118)

–   Face at most with yellow patterning, mesonotum not yellow-marked; if coxae darkened, not yellow-marked (differing in at least 1 other character) 7

7. Hind femora remarkably slender (Fig. 299); flagellum with about 70 segments; (ovipositor about 0.66 x basitarsus 3; femur 3 all black). *One UK record* .... *tibialis* Pfankuch (p. 119)

–   Hind femur of normal build (Fig. 300); flagellum with at most around 60 segments; (femur 3 usually predominantly red) 8

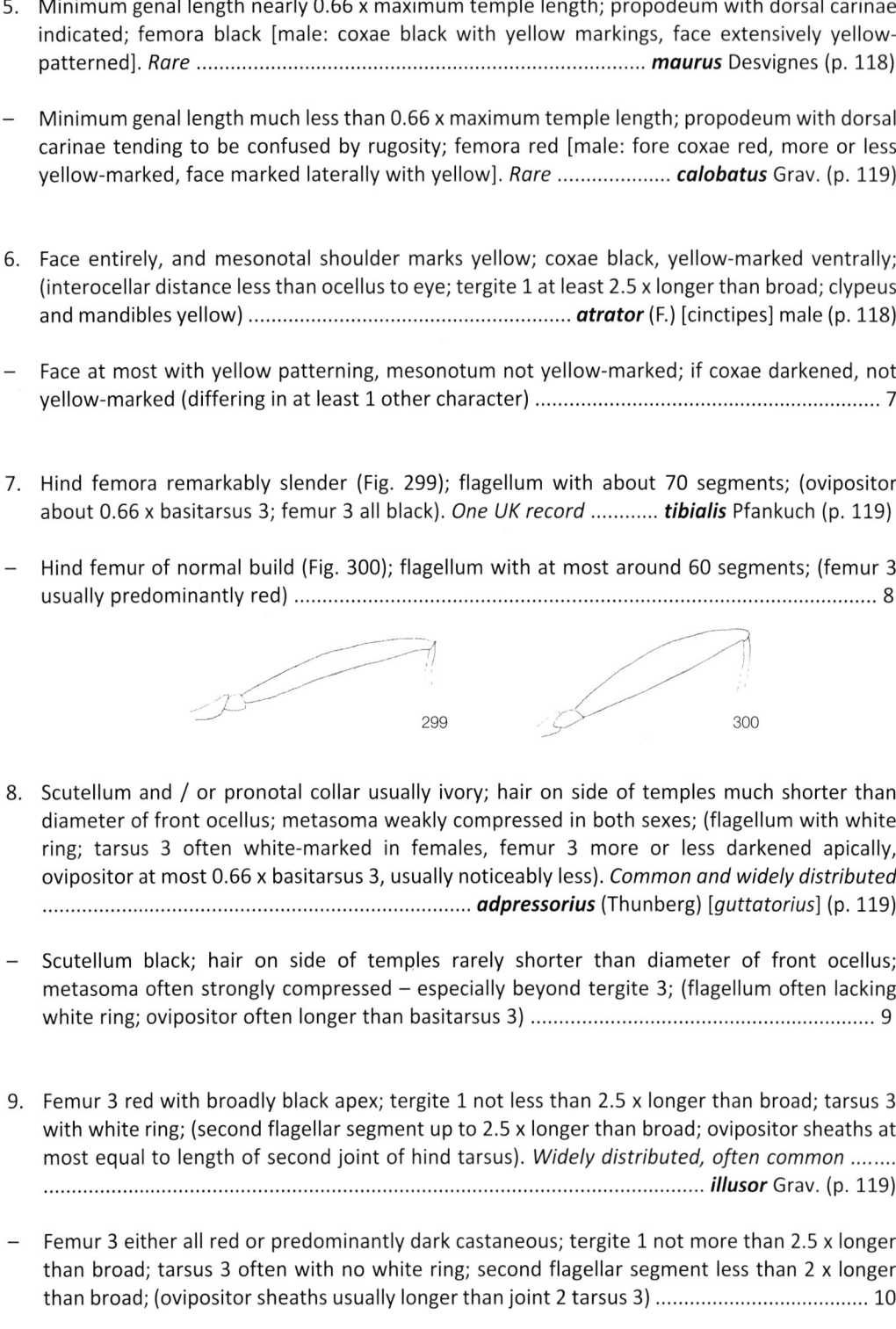

299   300

8. Scutellum and / or pronotal collar usually ivory; hair on side of temples much shorter than diameter of front ocellus; metasoma weakly compressed in both sexes; (flagellum with white ring; tarsus 3 often white-marked in females, femur 3 more or less darkened apically, ovipositor at most 0.66 x basitarsus 3, usually noticeably less). *Common and widely distributed* .... *adpressorius* (Thunberg) [*guttatorius*] (p. 119)

–   Scutellum black; hair on side of temples rarely shorter than diameter of front ocellus; metasoma often strongly compressed – especially beyond tergite 3; (flagellum often lacking white ring; ovipositor often longer than basitarsus 3) 9

9. Femur 3 red with broadly black apex; tergite 1 not less than 2.5 x longer than broad; tarsus 3 with white ring; (second flagellar segment up to 2.5 x longer than broad; ovipositor sheaths at most equal to length of second joint of hind tarsus). *Widely distributed, often common* .... *illusor* Grav. (p. 119)

–   Femur 3 either all red or predominantly dark castaneous; tergite 1 not more than 2.5 x longer than broad; tarsus 3 often with no white ring; second flagellar segment less than 2 x longer than broad; (ovipositor sheaths usually longer than joint 2 tarsus 3) 10

10. Ovipositor longer than basitarsus 3; female metasoma strongly laterally compressed; tergite 4 usually predominantly black; front and middle femora entirely reddish testaceous; (flagellum with second flagellar segment less than 2 x longer than broad). *Widespread, sometimes common* .... ***laevigator*** (Villers) (p. 120)

– Ovipositor much shorter than basitarsus 3; female metasoma not strongly compressed; tergite 4 usually red; front and middle femora either partly dark-suffused or entirely dark castaneous; (flagellum with second segment more than 2 x longer than broad). *Rare species* 11

11. Tergite 1 not more than 2.3 x longer than broad, female metasoma weakly compressed (Fig. 301); gena with large punctures (Fig. 302); face with dark pubescence; femora 1 and 2 red with dark markings, femur 3 red; (ovipositor shorter and broader). *Little known in the UK* .... ***femorator*** Desvignes (p. 120)

– Tergite 1 not less than 2.3 x longer than broad, female metasoma distinctly compressed (Fig. 303) gena with smaller punctures (Fig. 304); face with pale pubescence; femur 3 dark castaneous; (ovipositor longer and thinner). *Rare* .... ***nigripes*** Grav. (p. 120)

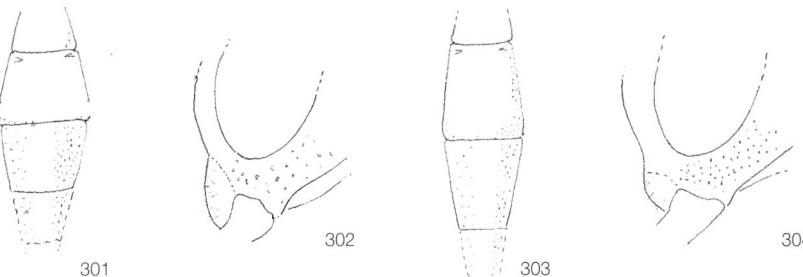

301   302   303   304

Species accounts for *Exetastes*

Exetastes fornicator (Fabricius, 1781)

Taxonomy: The Fabrician name *fornicator* comes from '*fornix*', a vault or arch – perhaps a reference to the posture of the metasoma during oviposition. The species is unlikely to be confused with any other *Exetastes*.

Distribution, abundance and phenology: *Exetastes fornicator* may possibly have been more numerous in the past, since very few records exist over the last century. Existing data are mostly (although not exclusively) from southern counties. *E. fornicator* has eight subspecies in the Nearctic region, these forming a geographic chain (Townes and Townes, 1978).

England: Many 'historical' records from Desvignes, Stephens, Marshall and Morley (none with data); England, Cornwall, Gurnard's Beach, viii.1936; others from: Cornwall, 1903, 'Dorset', Devon, 'Lydford, 1888'. [NHM]. More recent captures: Surrey, Worcester Park, 1949; Derbyshire, Kinrara, 1951 [NMS]. Yorkshire, Huggate [WE]; East Cornwall, Dannonchapel, Delabole, 2006. Ireland (O'Connor *et al*, 2007.): Tipperary (1927). Flight period: July and August.

Biology: *E. fornicator* has been reared from *Cucullia lactucae* (reputedly also *Lacanobia oleracea* (L.) on the European continent).

Exetastes atrator (Förster, 1771) [*cinctipes*]

Taxonomy: This is not a difficult species to recognise – although there may occasionally be some spreading of the black pigment such as to cause some degree of overlap with the next species.

Distribution, abundance and phenology: Morley (1908) considered this to be the commonest species of ichneumonid in the UK. However, there have been very few records over the last half-century or more.

England: a great many 'historical' records in NHM, also in HM and elsewhere. Scotland: one male, Edinburgh, in tea room of museum (thus possibly an 'import'), 16.vi.1990 [NMS]. Ireland (O'Connor *et al.*, 2007): formerly widespread. Flight period: June and July.

Biology: The known host is cabbage moth, *Mamestra brassicae* (L.). It seems possible that the decrease in domestic cabbage growing may be one reason for the apparent disappearance of a once abundant ichneumonid species. According to Morley (1908), the species also attacks *Lacanobia oleracea* (L.), which is not an exclusively garden insect. There are records of adults feeding on flowers of *Heracleum* and *Angelica* (Morley, 1908).

Exetastes illyricus Strobl, 1904

Taxonomy: *E. illyricus* was originally described from the Austrian mountains, thus would appear very unlikely to occur in southern England – particularly in urban London. Nevertheless, this determination is confirmed by comparison with the type. *E. illyricus* also occurs in the middle region of the European part of the former USSR (Kuslitsky, 1981).

Distribution, abundance and phenology: Recorded here as new to Britain on the basis of a few specimens taken in Sussex and in London.

England: one female, London, Lewisham, Sydenham Wood, garden edge Malaise trap, 20.vii-3.viii.1993 [HM]; one male, Sussex, South Ambersham, 18.viii.85 (A. Jones).

Biology: The host is unknown. As with the preceding (and certain other) black group members of the genus, there may perhaps be some link with garden environments.

Exetastes maurus Desvignes, 1856 Plate 15

Taxonomy: The name '*maurus*' (= Moorish) is a probable reference to the colouration of the species, which characteristic renders it quite distinct.

Distribution, abundance and phenology: There are some 'historical' specimens in the NHM collections, followed by a long gap, prior to the appearance of a small number of examples taken in and around gardens in the south of England. These localities include urban London.

England: Morley coll. (no data). Kent, Welling, Fanny on the Hill, 29.viii.1976. Kent, Barnehurst, five specimens, x.1976 [NHM]; Middlesex, Crouch End, x.1985, E. London Abney Park Cemetery, 14.ix.80; Lower Earley, Heath light trap in garden [NMS]. Middlesex, Bittern Hill; London, Poplar, Deptford; Surrey, Banstead Downs [HM]. Other recent records include one from London, in house, Woolwich, 28.ix.2006. Flight period: an autumnal species, late August-October.

Biology: The host is unknown, but seems very likely to be some 'horticultural' lepidopteran, bearing in mind the repeated association with urban / suburban locations. Aubert's (1978) listing of *Xestia xanthographa* (f. Seyrig, 1928) identifies a species which although common in gardens, is by no means restricted to same. *E. maurus* is sometimes attracted to light.

Exetastes calobatus Gravenhorst, 1829

Taxonomy: Easily distinguished from the preceding species on the basis of leg colour alone.

Distribution, abundance and phenology: Rare, although possibly less so during the 19th century. Records appertain to the southernmost counties of England.

England: Essex, Colchester; St Albans, 1871 (Marshall); several, Desvignes (no data) [NHM]. One female, Dorset, Broadstone Gardens, 22.ix.98 [NMS]; one male, Kent, Folkestone, Malaise trap, 8-28.ix.96 [HM]. Flight period: autumnal species, September.

Biology: One recent record implies a possible link with gardens.

Exetastes tibialis Pfankuch, 1921

Taxonomy: The elongate form of the hind femora is quite distinctive within the *Exetastes* 'red group'.

Distribution, abundance and phenology: Known in the UK from a single reared male specimen taken at Santon Downham in Norfolk (see below).

Biology: The sole British example of *E. tibialis* was reared from *Noctua orbona*. coll. 11.ii.2001 [NMS].

Exetastes adpressorius (Thunberg, 1822) [guttatorius]

Taxonomy: Perhaps the most easily recognised species in the red group, on the basis of colour pattern and form of the central metasomal tergites.

Distribution, abundance and phenology: Common and widely distributed. Adults often occur on Apiaceae. Ireland (O'Connor *et al*, 2007.): widespread. Flight period: mid-summer, July-August.

Biology: Valid host determinations for the species are: England: '? *Hoplodrina blanda*' (D. & S.). Spain: *Hoplodrina ambigua* (D. & S.) [NMS]. This is a species of more or less open country, including commons, hedgerows and similar habitats.

Exetastes illusor Gravenhorst, 1829

Taxonomy: *E. illusor* has been the subject of much confusion in the past, mainly due to the wide range of variation in colour. Northern examples tend to have the hind femora darkened to a greater or lesser extent. Two subspecies of *E. illusor* occur in the Nearctic region (Townes and Townes, 1978).

The species *E. geniculosus* has been separated on the basis mostly of inadequate colour characteristics. Males referred to this species have tergite 3 more transverse than in typical *illusor*. However, this character is much too variable in *illusor* to provide a definitive marker. According to Kuslitsky (1981), females differ in having the ovipositor about same length as fore tarsus segment 1 (segment 2 in *illusor*). However, these ratios cannot be confirmed, either for the single ('reputed') *geniculosus* female examined, or for *illusor* itself.

Distribution, abundance and phenology: Widespread and often common, extending to northern Scotland. The species is frequently taken on umbellifers. Ireland (O'Connor *et al.*, 2007): Down (det. Roman). Flight period: June to August.

Biology: I have encountered several reared examples from Colchester and Cambridgeshire, all lacking host data [NHM]. As with the previous species, habitats appear to be open country.

Exetastes laevigator (Villers, 1789)

Taxonomy: The long ovipositor readily distinguishes *laevigator* from the majority of red group *Exetastes* species.

Distribution, abundance and phenology: Relatively few specimens in older collections. England: Halling, W.K., plus a few others from Marshall – Desvignes [NHM]. More recently, the species has been taken in reasonable numbers in Malaise traps: Norfolk [NMS]; Surrey, Kent [HM]). Flight period: a mid-summer species, flying during June to August.

Biology: Unknown. Collecting data include fenland and Breckland.

Exetastes femorator Desvignes, 1856

Taxonomy: *E. femorator* has stood as one of the more obscure red group species in the past, owing to inadequate choice of taxonomic characters in the older literature.

Distribution, abundance and phenology: Known from a few specimens collected during the 19th century: two females, England: Kent, 'Deal, 1856' (no other data) – also lectotype female – 'Gt Britain' [NHM]. Only subsequent records: Yorkshire, Spurn, 17, 20, 23.vii.1948, (det. J. F. Perkins) [MU]. Ireland (O'Connor *et al.*, 2007): Wicklow (Stelfox – 'as Banchus'). The latter determination requires confirmation. Flight period: the limited existing data place the flight period in June and July.

Biology: The species is said to have the habit of running on sand, after the manner of a pompilid wasp (see Morley), an observation which clearly needs confirmation. These early data plus the few subsequent recent records seem to imply a link with sandy, coastal habitats. Host relationships are unknown.

Exetastes nigripes Gravenhorst, 1829

Taxonomy: Another species that has been difficult to determine from the older literature.

Distribution, abundance and phenology: Widely taken in earlier times, on the basis of material in older collections – mostly from southern England [NHM]. Later records are not numerous, but extend as far north as Inverness-shire.

Recent records: Wales, Glamorgan, Llangennith, Harding's Down, ex *Lacanobia oleracea* on *Pteridium aquilinum*, coll. 19.viii.2003. [NMS]. England: Yorkshire, Thrybergh CP, 1.viii.1984 [WE]. Ireland (O'Connor *et al.*, 2007): earlier records from Armagh; Down (1921, 29 – det. Johnson).

Biology: Habitats seem broadly similar to those of the commoner *Exetastes* species. It seems odd that the only valid host encountered (see above) is a common and widely distributed lepidopteran.

Genus *Banchus* Fabricius 1798

The genus *Banchus* is of predominantly Holarctic distribution. A total of 24 species are Nearctic (Townes and Townes, 1978), with 22 in the Palaearctic region (Fitton, 1985). Only 12 are found in the Western Palaearctic (Aubert, 1978), and 6 in the European part of the former USSR (Kuslitsky, 1981). On this basis, there would seem to be a higher proportion of species in the UK compared to Europe, than is usually expected for Banchinae.

Banchus species mostly inhabit open, shrubby country. According to Townes (1978), females fly low, males higher, in a weaving flight pattern. None have more than one generation a year, although see Fitton (1985) regarding European populations of *B. pictus*. In at least some species, the adults give off a strong pungent odour. 'Flag setae' may be found on the upper side of the postmedian segments of male antennae, their tips often being recurved.

As with *Exetastes*, several species of *Banchus* appear to have undergone a decline in numbers over the last half century or so. These are large, conspicuous insects, unlikely to have been overlooked by collectors.

Fore wing lengths of Banchus species range from 7-12 mm, usually towards the lower end of the stated range.

Key to species of *Banchus*

1. Maxillary palpi with fourth segment elongate and slender, at least equal to third, and with fifth segment minute in males (Fig. 305); tergites 2 *and* 3 with prominent thyridii (Fig. 306) 2

—   Maxillary palpi with segment 4 relatively broad, shorter than segment 3, the fifth segment large in males (Figs 307, 308, 309); tergites 2 and 3 with thyridii indistinct or absent 3

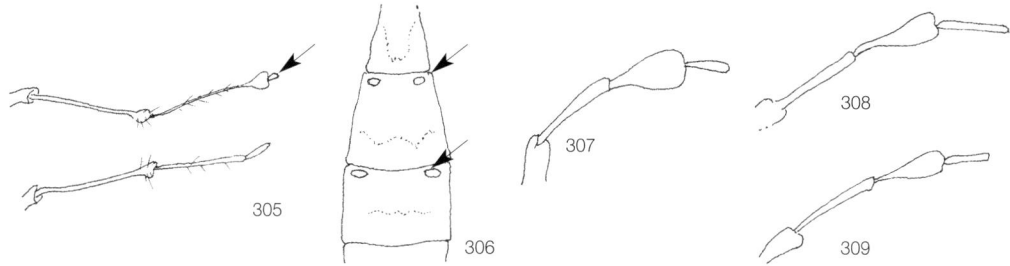

2. Scutellar spine longer than 0.8 x scutellum (Fig. 310); femur 3 broader (at most slightly greater than 6 x longer than broad), reddish-yellow in males, testaceous in female, rarely darkened ventrally; metasoma predominantly black – sometimes pale (rarely yellow) – margined on tergites 1-3 in males). *Rare, summer* .... ***palpalis*** Ruthe [*monileatus*] (p. 124)

—   Scutellar spine only about 0.3 x scutellum; femur 3 narrower (over 6.5 x longer than broad), black centrally, yellow-at base and apex (sometimes also dorsally); metasomal tergites black-and-yellow banded in both sexes. *Rare, early Spring* .... ***crefeldensis*** Ulbricht (p. 124)

3. **MALES** [hypopygium short – Fig. 311]; face yellow with black median stripe 4

– **FEMALES** [hypopygium long – Fig. 312]; face usually black in ground colour, with yellow markings laterally (excluding *pictus*) 9

MALES:

4. Flagellum with 'flag' setae present in distal part (these being erect, angled, and flattened, arising from a distinct groove) (Fig. 313) 5

– If 'flag' setae present, they are weakly developed, non-erect, and not arising from a groove 8

5. 3-4 flag setae per segment; metasoma distinctly compressed, tergites 2 and 3 largely yellowish and testaceous-marked, not black-banded; femur 3 all testaceous; scutellum with minute spine; (malar space at least 0.66 x mandible base (Fig. 314)). *Uncommon* ....
.... *falcatorius* (F.) (p. 124)

– 2 flag setae per segment; metasoma not noticeably compressed, the somites black-and-yellow banded; either femur 3 dark-marked or scutellar spine strong 6

6. Metasoma with tergite 1 having smoothly curved profile (Fig. 315); segment 4 of maxillary palpus very broad (Fig. 316); head and thorax with short hair – facial hairs at most somewhat greater than diameter of median ocellus; malar space a little greater than 0.5 x mandible base, face width (measured just above base of clypeus) not or little greater than height of eye; posterior face of propodeum usually situated at or before middle (Fig. 317); femur 3 entirely testaceous. *Common* .... *volutatorius* (L.) (p. 125)

– Metasoma with tergite 1 'humped' – i.e. with a distinct dorsal convexity lying approximately in line with the spiracles (Fig. 318) – in case of doubt, segment 4 of maxillary palpus not unusually broad; head and thorax with long hair – facial hairs up to about 2 x diameter of median ocellus; malar space at least 0.66 x mandible base (Fig. 319); face width greater than height of eye; posterior face of propodeum usually at or beyond middle; femur 3 partly blackish ventrally 7

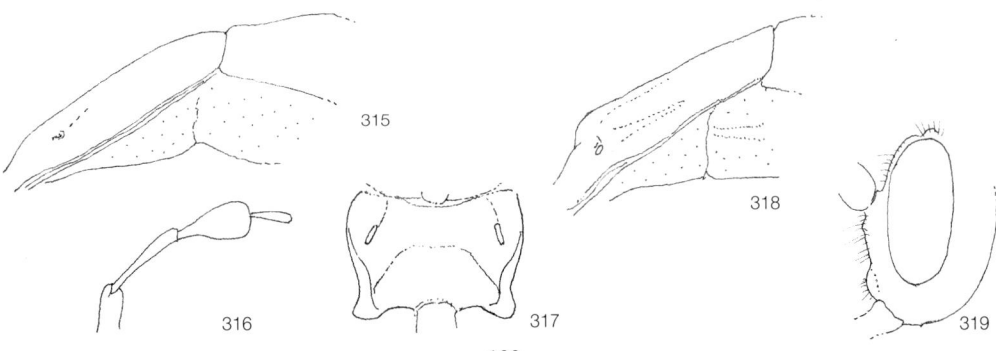

7. Scutellar spine minute to absent; flagellum testaceous, partly darkened. *Few recent records* .... ***dilatatorius*** (Thunberg) [*compressus*] (p. 125)

–   Scutellar spine strong; flagellum black, brownish distally. *One reputedly British specimen* .... ***moppiti*** Fitton (p. 125)

8. Metasoma predominantly blackish – posterior margin of tergites 1 and 2 (sometimes also 3) brownish-yellow – sometimes with a yellow and testaceous zone behind, scutellum predominantly black; segment 4 of maxillary palpus wider (Fig. 320). *Locally common in* Pinus *plantations* .... ***hastator*** (F.) (p. 125)

320

–   All metasomal tergites black and yellow-banded, scutellum yellow; segment 4 of maxillary palpus narrower (Fig. 321). *Rare* .... ***pictus*** F. (p. 126)

321

FEMALES:

9. Metasoma with tergite 1 usually distinctly 'humped' – i.e. with a distinct dorsal prominence lying approximately in line with the spiracles (Fig. 322); (tergites either predominantly dark, or else black-and-yellow banded) 10

–   Metasoma with tergite 1 usually with smoothly curved profile (as Fig. 323) 12

322

323

10. Tergites 5 and 6 broader than long, only moderately compressed (Fig. 324); face width at most equal to height of eye; metasomal tergites predominantly dark, often with brownish and / or yellowish hind margins; sternites more or less pale-margined in front (sometimes also behind) .... ***hastator*** (F.) (p. 125)

–   Tergites 5-7 longer than broad, with knife-like compression (tergite 4 parallel-sided) (Fig. 325); face width greater than height of eye; metasomal sternites only pale-marked behind (if tergites not black-and-yellow banded, cf. *falcatorius*) 11

324

11. Scutellar spine minute or absent; flagellum testaceous, partly darkened; tergite 1 not more than 1.4 x longer than broad .... ***dilatatorius*** (Thunberg) [*compressus*] (p. 125)

–   Scutellar spine strong; flagellum blackish distally; tergite 1 not less than 1.6 x longer than broad .... ***moppiti*** Fitton (p. 125)

325

12. Metasoma elongate, more strongly compressed – tergite 7 not less than 3 x longer than broad, tergites 2 and 3 predominantly testaceous, with some degree of piceous suffusion; scutellum spine minute .... *falcatorius* (F.) (p. 124)

   – Metasoma less elongate, more weakly compressed – tergite 7 much less than 3 x longer than broad, tergites all black or black-and yellow banded; (scutellum spine often strong) 13

13. Hind femur entirely testaceous, scutellum and metasoma black; tergite 7 broader than long; posterior face of propodeum at or before middle .... *volutatorius* (L.) (p. 125)

   – Hind femur with greater or lesser degree of dark suffusion, scutellum yellow-marked, metasoma black-and-yellow banded; tergite 7 longer than broad; posterior face of propodeum at or beyond middle .... *pictus* F. (p. 126)

Species accounts for *Banchus*

Banchus palpalis Ruthe, 1859 [*monileatus*]

Taxonomy: Given due attention to characters of the palpi, the two species forming the *palpalis* group are easily separated, both from one another and from the remainder of the genus.

Distribution, abundance and phenology: A few specimens exist in older collections, including material from Ireland (see Fitton, 1985). More recently, examples have been taken in Scotland: one male, Angus, Glen Clova, 9.vii.1995 [NMS]; also England: Yorkshire, Balder Head, 12.vii.2012 [WE]. Ireland (O'Connor *et al.*, 2007): Donegal (1917); Down (1927); Dublin (1941). Flight period: early to mid-summer.

Biology: Habitats appear generally to be upland localities.

Banchus crefeldensis Ulbricht, 1916

Taxonomy: The present is unlikely to be mistaken for any other *Banchus* species.

Distribution, abundance and phenology: According to Fitton (1985), *B. crefeldensis* is restricted to the Scottish Highlands, so far as the British mainland is concerned. Ireland: Stelfox records males flying around Ivy (*Hedera*) and Gorse (*Ulex*) (Fitton, 1985). More recently, Scotland: one male, East Ross, 13.iv.1992 [NMS]. Flight period: early in the year (February-May), which may be one reason for its apparent rarity.

Biology: *B. crefeldensis* appears to occur at high altitudes. The host is *Aporophylla lutulenta* (D. & S.) – including form *luneburgensis* (Fitton).

Banchus falcatorius (Fabricius, 1804)

Taxonomy: The strongly compressed female metasoma, together with predominantly testaceous central segments render this species distinctive.

Distribution, abundance and phenology: There are many specimens in collections, pre-1950s. More recently, England: Norfolk, Santon Downham, Malaise trap, 1983-5; Norfolk, East Wretham, ex *Agrotis exclamationis* (L.) coll. 29.viii.; Berkshire, Silwood Park, Malaise trap [NMS]; others in York and Manchester Museums. Ireland (O'Connor *et al.*, 2007): 'status unconfirmed'. Flight period: predominantly early June to mid-August (Fitton, 1985).

Biology: Known hosts include *Agrotis segetum* (D. & S.), given by Fitton (1985) as being the most reliable amongst existing records. Additionally, *Agrotis exclamationis* is a verified host. Both noctuids are common and widely distributed in the UK. Habitats are open country.

Banchus volutatorius (Linnaeus, 1758)

Taxonomy: Males have the familiar black-and-yellow banding found in this sub-group, while females have the metasoma entirely black.

Distribution, abundance and phenology: *B. volutatorius* is by far the commonest and most widely distributed *Banchus* species. Probably found in most parts of the UK (although seemingly with much local variation in frequency). Ireland (O'Connor *et al.*, 2007): widespread. Flight period: mostly taken between mid-June and early August.

Biology: Confirmed hosts include: *Anarta myrtilli* (L.), *Mamestra brassicae* (L.), *Lacanobia oleracea, Hadena,* and *Xestia* species. *B. volutatorius* may be found on open country generally, including heaths, downland, commons, and amongst scrub. Adults are sometimes found on the flower heads of Apiaceae.

Banchus dilatatorius (Thunberg, 1822) [*compressus*]

Taxonomy: The unusually wide face is distinctive within the genus.

Distribution, abundance and phenology: Many specimens in the National Collection in London. Also Scotland: one female, Fife, St Andrews, 6.iv.1949 [NMS]. Flight period: late March to late April, the species has been collected on *Salix* catkins. As with *B. crefeldensis*, the early appearance may account in part for its apparent rarity.

Biology: Fitton cites several reputed host records – adding that he has encountered none personally.

Banchus moppiti Fitton, 1985

Taxonomy: Closely related to the previous species.

Distribution, abundance and phenology: One reputedly British specimen of 'historical' status (no data). Flight period: February to April (following records from the continent of Europe).

Biology: Unknown.

Banchus hastator (Fabricius, 1793)

Taxonomy: The species is quite isolated within the genus. Its divergent metasomal colouration is a distinctive recognition feature for both sexes. *B. hastator* is regarded by Townes and Townes (1978) as being related to *crefeldensis* (although with no data given in support of this statement).

Distribution, abundance and phenology: Widespread in *Pinus* plantations in Scotland, also taken in England: Norfolk, Santon Downham [NMS]; Yorkshire, Ogden Clough; Netherton [WE]. Ireland (O'Connor): Wicklow. Flight period: May, through to early July.

Biology: The host is *Panolis flammea* (D. & S.). Fitton (1985) lists other 'in lit.' hosts, but casts doubt upon their authenticity. Clearly, the species occurs in coniferous forest biotopes.

Banchus pictus Fabricius, 1798, Plate 16 & cover photograph

Taxonomy: Fitton (1985) discusses the possibility of two species being involved under this name. However, evidence was regarded as being inconclusive, and a partial second generation of a single species was given as an alternative explanation.

Distribution, abundance and phenology: Some 'historical' data [NHM]; Ireland (O'Connor *et al.*, 2007): early Irish records stated to be erroneous. More recently, there are scattered records to as far north as Cheshire: England: Hertfordshire, Bricket Wood, 1930s; Hampshire, 1981; north Staffordshire, 23.v.1917; Shropshire, Market Drayton; Hampshire, Forest of Bere, 17.iv.81 [NHM]; Cheshire, Abbots Moss, flying over *Calluna* 7.v.1976; Norfolk, Santon Downham, Malaise trap 21-31.v-1985 [NMS]; (no data), Moore Coll. [HM]. Wales: Powys, Presteigne, 8-9.v.1953 [NHM]. Flight period: another species with an early flight period: April to May. *Banchus pictus* is apparently univoltine in the UK.

Biology: I have encountered one 'probably authentic' rearing: Hampshire, New Forest, ex *Agrochola helvola* (L.), 1913 (Lyle) [NHM]. The recorded host is ubiquitous in its ecology. Fitton cites other reputed hosts from the literature.

Genus *Rynchobanchus* Kriechbaumer, 1894

Species account for *Rynchobanchus*

Rynchobanchus flavopictus Heinrich, 1937 Plate 17

Taxonomy: *R. flavopictus* is unlikely to be mistaken for any other British banchine.

Distribution, abundance and phenology: Three British specimens known, all collected during comparatively recent times: female, England: Berkshire, Hermitage, Fence Wood, ex cocoon, 1981. Male / female, S. Hampshire, Austin Copse, 29.iv.2011 [NHM]. Flight period: early spring (perhaps April to May).

Biology: No records exist for host preference.

Appendix

Hosts and parasitoids

The following table summarises the known host-parasitoid relationships as detailed in the main body of this work. Host relationships are still poorly known in Banchinae (as with most other ichneumonids). To underline this fact, it is only necessary to point out that the commonest British banchine, *Lissonota coracina*, has never been reared in this country – and its host relationships are known from a single specimen reared from '*Crambus* sp.' in the USA. Conclusions drawn from the information given therefore have to lie very much in the 'provisional' category, and can form no more than a basis for a deeper understanding of parasitoid-host relationships in the subfamily. Likewise, while the table may provide a rough guide as to the identification of a banchine species reared from a known host – this cannot possibly take the place of taxonomic diagnosis. I have usually excluded reputed hosts from the table – exceptions being where a given relationship might usefully be followed up by further hosts rearing.

Table 1. Host-parasitoid relationships

Hosts	Parasitoids
ADELIDAE	
Nemophora fasciella	*Lissonota consobrina*
PSYCHIDAE	
Diplodoma laichartingella	*Lissonota saturator, L. nigridens*
Epichnopteryx plumella	*Lissonota linearis*
Luffia fershaultella	*Lissonota luffiator*
Proutia betulina	*Lissonota luffiator*
Whittleia retiella	*Lissonota linearis*
'*Psyche intermediella*'	*Lissonota nigridens*
TINEIDAE	
Infurcitinea albicomella	*Lissonota picticoxis*
Nemapogon cloacella	*Lissonota punctiventris*
Nemaxera betulinella	*Lissonota punctiventris*
Triaxomera fulvimetrella	*Lissonota punctiventris*
Triaxomera parasitella	*Lissonota palpator*
COSSIDAE	
Cossus cossus	*Lissonota setosa*
SESIIDAE	
Pyropteron muscaeformis	*Lissonota pimplator*
Sesia ?crabroniformis (or) apiformis	*Lissonota fulvipes*
Synanthedon andrenaeformis	*Lissonota freyi, L. canaliculata*
Synanthedon culiciformis	*Lissonota frontalis*
Synanthedon formicaeformis	*Lissonota frontalis*
Synanthedon tipuliformis	*Lissonota nitida*
Synanthedon scoliaeformis	*Lissonota plana*

Hosts	Parasitoids
CHOREUTIDAE	
Anthophila fabriciana	*Lissonota stigmator*
TORTRICIDAE	
PHALONIINAE	
Aethes margaritana	*Glypta longispinis*
Agapeta zoegana	*Glypta sculpturata*
? Cochylis hybridella	*Glypta mensurator*
Commophila aeneana	*Glypta ulbrichti*
'*Eupoecilia notulana*'	*Glypta rufata*
TORTRICINAE	
Acleris ferrugana	*Glypta pictipes*
Acleris hastiana	*Glypta parvicaudata*
'*? Acleris variegana*'	*Apophua bipunctoria*
Adoxophyes orana	*Apophua bipunctoria*
Aphelia paleana	*Diblastomorpha cylindrator, D. rostrata*
Aphelia unitana	*Diblastomorpha cylindrator*
'*Aphelia viburnana*'	*Apophua genalis*
Apotomis betuletana	*Glypta consimilis*
? Apotomis betuletana	*Apophua bipunctoria*
Apotomis capreana	*Glypta nigrina*
'*? Apotomis sauciana*'	*Glypta consimilis*
Archips (*Cacoecia*) species	*Telutaea brischkei*
Clepsis spectrana	*Conoblasta monoceros*
? Bactra lancealana	*Conoblasta elongata*
Cnephasia lacunana	*Conoblasta ceratites*
Ditula angustiorana	*Conoblasta lapponica*
Endothenia gentianaeana	*Conoblasta woerzi*
Endothenia nigricostana	*Conoblasta paludosa, C. elongata*
'*? Epiphyas postvittana*'	*Apophua bipunctoria*
Pandemis cerasana	*Apophua bipunctoria, Conoblasta lapponica, Glypta nigrina*
P. cinnamomeana	*Apophua bipunctoria, A. evanescens*
Pandemis corylana	*Apophua evanescens, Glypta nigrina*
Ptycholoma lecheana	*Apophua bipunctoria*
Tortrix viridana	*Glypta nigrina, Lissonota mutator*
OLETHREUTINAE	
Anacampsis blattariella	*Glypta consimilis*
? Pseudococcyx	*Glypta resinanae*
Cydia grunetiana	*Glypta tenuicornis*
Cydia nigricana	*Glypta haesitator*
Cydia splendana	*Lissonota carbonaria*
Cydia succedana (D. & S.) (*ulicetana* (Haw.))	*Glypta trochanterata*
Dichrorampha sp.	*Lissonota serena*
Dichrorampha senectana	*Glypta femorator*

Hosts	Parasitoids
Dichrorampha simpliciana	*Glypta bifoveolata, Glypta incisa*
Epiblema cirsiana	*Glypta similis*
Eucosma foenella	*Glypta incisa*
? Eucosma hohenwartiana	*Glypta vulnerator*
Epiblema rosaecolana	*Glypta haesitator*
Epiblema scutulana	*Glypta similis*
Epinotia brunnichana	*Lissonota folii*
Epinotia solandriana	*Glypta nigrina*
Epinotia sordidana	*Lissonota folii*
Epinotia tedella	*Lissonota dubia*
Eucosma conterminana	*Glypta microcera*
?Grapholita molesta	*? Lissonota gracilenta*
Gypsonoma aceriana	*Lissonota lineata*
Olethreutes mygindiana	*Glypta fronticornis*
Retinia resinella	*Lissonota carbonaria*
Rhyacionia	*Glypta resinanae, Lissonota carbonaria*
? Spilonota ocellana	*? Glypta scutellaris*
(?) - do -	*? Glypta haesitator*
Zeirephera griseana	*Lissonota folii*

YPSOLOPHIDAE

Ypsolopha dentella	*Lissonota quadrinotata*
Ypsolopha horridella	*Lissonota gracilipes*
Ypsolopha parenthesella	*Lissonota palpalis, L. buccator, L. gracilipes*
Ypsolopha ustella	*Lissonota palpalis, L. gracilipes*

OECOPHORIDAE sensu lato

Agonopterix conterminella	*Conoblasta ceratites, Glypta consimilis*
Agonopterix liturosa	*? Lissonota fletcheri*
Agonopterix nervosa	*Conoblasta ceratites*
Agonopterix ulicitella	*Conoblasta ceratites*
Aloboria geoffrella	*Lissonota simulator, L. semirufa*
Aplota palpella	*Lissonota tenerrima*
Esperia sulphurella	*Lissonota biguttata, ? L. semirufa, L. arborator*
	L. simulator
Schiffermuelleria grandis	*Lissonota arborator*

GELECHIIDAE

Anarsia spartiella	*Conoblasta ceratites*
? Mirificarma lentiginosella	*Lissonota fletcheri*

PYRALIDAE

Anerastia lotella	*Lissonota lineata*
Dioryctria abietella	*Lissonota dormitor*
Epischnia bankesiella	*Syzeuctus furcator*
Oncocera genistella	*Syzeuctus furcator*
Pempelia palumbella	*Syzeuctus furcator*
Synaphe punctalis	*Lissonota cruentator*

Hosts	Parasitoids
CRAMBIDAE	
Eudonia lacustrata [??Dipleurina]	*Lissonota tenerrima, L. versicolor*
Eudonia lineola	*Lissonota versicolor, L. pleuralis*
Eudonia truncicolella	*Lissonota tenerrima*
Mecyna asinalis	*Lissonota subaciculata*
GEOMETRIDAE	
LARENTIINAE	
? Anticlea badiata	*Cryptopimpla caligata*
Camptogramma bilineata	*Cryptopimpla calceolata*
Entephria caesiata	*Cryptopimpla anomala, C. hertrichi*
Entephria flavicinctata	*Cryptopimpla hertrichi, C. errabunda*
Eppirhoe alternata	*Cryptopimpla errabunda*
Eulithis testata	*Cryptopimpla anomala, C. errabunda*
Operophtera brumata	*Lissonota biguttata*
ENNOMINAE	
Alcis repandata	*Lissonota biguttata*
Ligdia adusta	*Cryptopimpla arvicola*
NOCTUIDAE	
NOCTUINAE	
Agrotis exclamationis	*Banchus falcatorius*
Agrotis segetum	*Banchus falcatorius*
Eugnorisma glareosa	*Arenetra pilosella*
Xestia	*Banchus volutatorius*
HADENINAE	
Anarta myrtilli	*Banchus volutatorius*
Cerapteryx graminis	*Arenetra pilosella*
Hadena	*Banchus volutatorius*
Lacanobia oleracea	*Exetastes fornicator, E. atrator, E. illusor,*
	E. nigripes, Banchus volutatorius
Mamestra brassicae	*Exetastes atrator, Banchus volutatorius*
'*? Orthosia stabilis*'	*Alloplasta piceator, A. plantaria*
O. gothica	*Alloplasta piceator*
O. gracilis	*Alloplasta piceator*
O. miniosa	*Alloplasta piceator*
Panolis flammea	*Banchus hastator*
CUCULLIINAE	
Agrochola helvola	*Banchus pictus*
Aporophylla lutulenta	*Banchus crefeldensis*

Hosts	Parasitoids

AMPHIPYRINAE

Apamea crenata — Lissonota lineolaris
Lissonota clypeator

Gortyna flavago — Lissonota digestor
Hoplodrina ambigua — Exetastes adpressorius
'? Hoplodrina blanda' — ? Exetastes adpressorius
Luperina nickerlii — Lissonota impresso, L. sabulosa
Mesapamea secalis — Lissonota fundator, L. clypeator
Oligia sp. — Lissonota clypeator

References and further reading

Aubert, J.F. 1978. *Les Ichneumonides ouest-palearctiques et leurs hotes 2. Banchinae et Suppl. aux Pimplinae*. Laboratoire d'Evolution des Etres Organises, Paris & EDIFAT-OPIDA, Echauffour.

Aubert, J.F. 1993. Nouvelles precisions sur la systematique de quelques Ichneumonides (Hymenoptera, Ichneumonidae). *Nouvelle Revue d'Entomologie*. **10**(3):211-221.

Beirne, B. P. 1941. A consideration of the cephalic structures and spiracles of the final instar larvae of the Ichneumonidae (Hym.). *Transactions of the Society of British Entomology*. **7**: 123-190.

Bridgman, J. B. 1882 (July). Further additions to Mr. Marshall's catalogue of British Ichneumonidae. *Transactions of the Entomological Society of London*. **1882**: 141-164.

Bridgman, J. B. 1886 (October). Further additions to Mr. Marshall's catalogue of British Ichneumonidae. *Transactions of the Entomological Society of London*. **1886**: 335-373.

Bridgman, J. B. 1887 (December). Further additions to Mr. Marshall's catalogue of British Ichneumonidae. *Transactions of the Entomological Society of London*. **1887**: 361-379.

Bridgman, J. B. 1889 (October). Further additions to Mr. Marshall's catalogue of British Ichneumonidae. *Transactions of the Entomological Society of London. **1889**: 09-439.

Bridgman, J. B. 1890 (July). Notes on Hymenoptera in the neighbourhood of Norwich; and on the genus Glypta, Gr.. *Transactions of the Norfolk and Norwich Naturalists' Society*. **5**: 61-72

Dasch, C. E. 1988. Ichneumon-flies of America north of Mexico: 10. Subfamily Banchinae, tribe Glyptini. *Memoirs of the American Entomological Institute*. No **43**.

Desvignes, T. 1856. *Catalogue of British Ichneumonidae in the collection of the British Museum*. London.

Fitton, M. G. 1976. The Western Palaearctic Ichneumonidae (Hymenoptera) of British authors. *Bulletin of the British Museum (Natural History), Entomology series*. **32**: 301-373.

Fitton, M. G. 1982. A catalogue and reclassification of the Ichneumonidae (Hymenoptera) described by C. G. Thomson. *Bulletin of the British Museum (Natural History), Entomology series*. **45**(1): 1-119.

Fitton, M G. 1985. The ichneumon-fly genus Banchus (Hymenoptera) in the Old World. *Bulletin of the British Museum (Natural History), Entomology series*. **51**(1): 1-60.

Fitton, M. G., Shaw, M. R. and Gauld, I.D. 1988. Pimpline Ichneumon-flies. Hymenoptera, Ichneumonidae (Pimplinae). *Handbooks for the Identification of British Insects* **7**(i).

Gravenhorst, J. L. C. 1829. *Ichneumonologia Europaea (Pars 1-3)*. Vratislaviae.

Holmgren, A. E. 1860. Försök till uppställning och beskrifning af Sveriges Ichneumonider. Tredje Serien. Fam. Pimplariae. (Monographia Pimplararium Sueciae). *Kongliga Svenska Vetenskapsakademiens Handlingar*. (B). **3**(10): 1-76.

Horstmann, K. 2003. Revisionen von Schlupfwespen-arten VII (Hymenoptera Ichneumonidae). *Mitt. Munch. Ent. Ges.* **93**: 25-37.

Kuslitsky, 1981. *In* Kasparyan (ed.) *A guide to the insects of the European part of the USSR. Hymenoptera, Ichneumonidae. Subfamily Banchinae. Opredeliteli' Faune SSSR.* **3**: 276-316. [in Russian]

Morley, C. 1908. *Ichneumonologia Brittanica, iii.* The ichneumons of Great Britain. Pimplinae. London.

O'Connor, J. P., Nash, R., and Fitton, M. G. 2007. *A Catalogue of the Irish Ichneumonidae (Hymenoptera: Ichneumonoidea). Occasional Publications of the Irish Biological Society.*

Perkins, J. F. 1959. Hymenoptera. Ichneumonoidea. Ichneumonidae, key to subfamilies and Ichneumoninae-1. *Handbooks for the Identification of British Insects* **7**(2ai), 1-116.

Rey del Castillo, C. 1992. Revision de la especies oeste-palearticas del subgenero Loxonota Aubert, 1978 (Hymenoptera: Ichneumonidae). *Annales de la Societe Entomologique de France. (N.S.).* **28**: 133-156.

Short, J. R. T. 1978. The final larval instars of the Ichneumonidae. *Memoirs of the American Entomological Institute.* No **25**.

Thomson, C. G. 1877. XXVIII. Bidrag till kännendom om Sveriges Pimpler. *Opuscula Entomologica. Lund.* **VIII**: 732-777.

Thomson, C. G. 1889a. XXXIX. Öfversigt af arterna inom slägtet Glypta (Grav.). *Opuscula Entomologica. Lund.* **XIII**: 1321-1353.

Thomson, C. G. 1889b. XLI. Bidrag till Sveriges insectfauna. *Opuscula Entomologica. Lund.* **XIII**: 1401-1438.

Townes, H. T. 1969. Genera of Ichneumonidae, Part 3 (Lycorininae, Banchinae, Scolobatinae, Porizontinae). *Memoirs of the American Entomological Institute.* **13**: 1-307.

Townes, H. T., and Townes, M. 1978. Ichneumon-flies of America north of Mexico: 7. Subfamily Banchinae, tribes Lissonotini and Banchini. *Memoirs of the American Entomological Institute.* **26**: 1-614.

Wahl, D. B. 1988. *A review of the mature larvae of the Banchini and their phylogenetic significance, with comments on the Stilbopinae (Hymenoptera:Ichneumonidae). In* Gupta, V. K. (ed.) *Advances in Parasitic Hymenoptera Research*, pp.147-161. Leiden, New York, Copenhagen, Cologne.

Yu, D. S. and Horstmann, K. 1997. A catalogue of world Ichneumonidae (Hymenoptera). *Memoirs of the American Entomological Institute.* **58**: 1-1558.

Index

Index to Tribe, genera and species. Main entries and start of sections are given in **bold**. Synonyms are in *italic*, as are species erroneously recorded as British / Irish.

abundans, Diblastomorpha 16
accusator, Lissonota 20, 96, **102**
acuminator, Banchus 23
admontensis, Lissonota 20, 78, **79**
adpressorius, Exetastes 22, 116, **119**
affinis, Lissonota 20, 21
agnata, Lissonota 19
albifrons, Apophua 16
albitarsoria, Alloplasta 18
albitarsus, Alloplasta 18
albobarbata, Lissonota 20
albopictor, Exetastes 22
albopictus, Exetastes 22
algerica, Glypta 17
Alloplasta 18, 55, **57**
alpina, Lissonota 20
alpina, Glypta 16
alpinus, Exetastes 23
alticola, Banchus 24
altipes, Cryptopimpla 22, 110, **112**, 149
annulata, Glypta 16
annulatus, Exetastes 23
anomala, Cryptopimpla 22, 109, **111**
antennalis, Lissonota 20, 83, **86**
Apophua 16, **28**, 30
arborator, Lissonota 2, 97, **104**
areolaris, Glypta 16
areolata, Lissonota 21
argiola, Lissonota 20, 80, **81**
aries, Banchus 23
arreptans, Glypta 18
artemesiae, Lissonota 20
arvicola, Cryptopimpla 22, 108, **110**
assimilis, Lissonota 21
atrator, Exetastes 23, 115, 116, **118**
ATROPHINI 8, 13, 18, 25, **53**, 54
australis, Glypta 17

baltica, Apophua 16
BANCHINI 7, 13, 22, 25, **113**, 114
Banchus 23, 114, **121**
basalis, Lissonota 20, 22
bellator, Lissonota 21
bellatrix, Lissonota 21
benoisti, Exetastes 23
berolinae, Glypta 17

bicolor, Syzeuctus 18
bicornis, Diblastomorpha 16
bicornis, Syzeuctus 18, **55**
bifoveolata, Glypta 17, 47, **48**
biguttata, Lissonota 19, 59, **63**, 75
bipunctatus, Banchus 24
bipunctoria, Apophua 16, 29, **30**
blanda, Cryptopimpla 22
brachycentra, Cryptopimpla 22
brevicornis, Glypta 17
brischkei, Glypta 17
brischkei, Telutaea 16, **28**, 143
?buccator, Lissonota 20, **106**

calcaratus, Banchus 24
calceolata, Cryptopimpla 22, 109, **111**
caligata, Cryptopimpla 22, 109, **110**
calobates, Exetastes 23
calobatus, Exetastes 23, 116, **119**
Campocineta 20, 59, **75**, 78
canaliculata, Lissonota 19, 62, **68**
carbonaria, Lissonota 20, 91, 97, **102**
carinifrons, Lissonota 22
catenator, Lissonata 19
caudata, Lissonota 20
ceratites, Glypta 16, 35, **38**
certator, Banchus 24
cicatricosa, Apophua 16, 29, **30**, 143
cinctipes, Exetastes 23
clavator, Exetastes 23
clypealis, Lissonota 20, **82**
clypeator, Lissonota 19, 69, **71**, 148
commixta, Lissonota 21
compressus, Banchus 23
Conoblasta 16, 28, **32**
consimilis, Glypta 17, **40**
consobrina, Lissonota 20, 85, **87**
coracina, Lissonota 21, 80, **82**
corniculata, Diblastomorpha 16
cothurnatus, Exetastes 23
coxator, Lissonota 22
crassipes, Lissonota 19
creditor, Alloplasta 18
crefeldensis, Banchus 23, 121, **124**
crenulata, Apophua 16
croaticus, Exetastes 23

cruenta, Lissonota 21
cruentator, Lissonota 20, **73**
cruentatrix , Lissonota 20
cryptator, Lissonota 19
Cryptopimpla 22, 55, **108**
cubitoria, Apophua 16
culiciformis, Lissonota 21, 92, **93**
cultratus, Banchus 24
curvicoxa, Glypta 17
cylindrator, Diblastomorpha 16, **31**, 144
cylindrator, Lissonota 19
cylindratrix, Diblastomorpha 16

deversor, Lissonota 19, 61, **67**
Diblastomorpha 16, **31**
digestor, Lissonota 19, 69, **71**
dilatatorius, Banchus 23, 123, **125**
dioszhegyi, Lissonota 20
dormitor, Lissonota 19, 62, **65**
dubia, Lissonota 21, **94**
duplanae, Lissonota 21

elegans, Diblastomorpha 16
elegantula, Glypta 17
elongata, Glypta 16, 33, **36**
enervator, Lissonota 19
ephippigera, Diblastomorpha 16
errabunda, Cryptopimpla 22, 109, **111**
?errabunda, Lissonota 21
erythrina, Lissonota 21, 75, **78**
erythrogaster, Diblastomorpha 16
evanescens, Apophua 16, 29, **30**
exareolata, Lissonota 21
excavator, Lissonota 19
eximia, Lissonota 20
expansor, Exetastes 23
Exetastes 22, **114**
extincta, Glypta 16, 34, **37**

facialis, Exetastes 23
facialis, Lissonota 19
falcator, Banchus 24
falcatorius, Banchus 23, 122, **124**
farrani, Banchus 24
femoralis, Banchus 24
femorata, Lissonota 19
femorator, Exetastes 23, 117, **120**
femorator, Glypta 17, 43, **45**
femoratrix, Glypta 17
filicornis, Glypta 17
flavipes, Glypta 17

flavipes, Lissonota 19
flavolineata, Apophua 16
flavopictus, Rynchobanchus 24, **126**, 150
fletcheri, Lissonota 21, 77, 84, **87**
folii, Lissonota 21, 98, **104**
formosa, Lissonota 22
fornicator, Exetastes 23, 115, **117**
freyi, Lissonota 19, 62, **67**, 147
frontalis, Lissonota 19, 62, **67**
fronticornis, Glypta 16, 34, **37**
fulvipes, Lissonota 19, 61, **65**, 147
fundator, Lissonota 20, 69, **70**
fuscator, Syzeuctus 18, 55, **56**, 145

genalis, Apophua 16, 29, **30**
genator, Lissonota 21, **78**
genucincta, Alloplasta 18
?geniculosus, Exetastes 23
gladiator, Lissonota 19
Glypta 16, 17, 27, 28, **38**, 41
GLYPTINI 7, 13, 16, 25, **26**
gracilenta, Lissonota 21, 91, **92**
gracilipes, Lissonota 21, 82, **83**
gracilis, Glypta 18
guttatorius, Exetastes 22
guttifer, Exetastes 22

haesitator, Glypta 17, 43, **46**
haesitatrix, Glypta 17
halidayi, Lissonota 21, 91, **92**
hastator, Banchus 24, 123, **126**
hertrichi, Cryptopimpla 22, 109, **112**
heydeni, Glypta 17
hians, Lissonata 19
histrio, Banchus 23
histrio, Lissonota 20, 73, **74**, 148
humeralis, Lissonota 20
humerella, Lissonota 20

illusor, Exetastes 23, 116, **119**
illyricus, Exetastes 23, 115, **118**
impressor, Lissonota 19, 20, 69, **70**
inareolata, Lissonota 21
incerta, Lissonota 21
incisa, Glypta 17, 47, **48**
incurvator, Exetastes 23
insignita, Lissonota 20
intersectus, Banchus 23
irrigua, Lissonota 21
irrisorius, Syzeuctus 22

jaroslavensis, Glypta 17
jugorum, Lissonota 21
junci, Exetastes 23

kaisdii, Cryptopimpla 22
kolosovi, Banchus 24

labiatus, Banchus 23
laevigator, Exetastes 23, 117, **120**
Lampronota 19, 59, **60**
lapponica, Glypta 16, 35, **37**
lapponica, Lissonota 21
lata, Alloplasta 18
lateralis, Lissonota 21
lavrovi, Banchus 24
leptogaster, Cryptopimpla 22
leucogona, Lissonota 22
levigator, Exetastes 23
linearis, Lissonota 21, 76, **79**
lineata, Glypta 17, 42, **44**
lineata, Lissonota 20, 73, **74**
lineolaris, Lissonota 19, 58, **60**, 146
Lissonota 19, 55, **58**, 59, 68
lissonotoides, Lissonota 19
longicauda, Glypta 17
longispinis, Glypta 17, 43, **45**
Loxonota 20, 58, **68**, 72
luffiator, Lissonota 21, 84, **86**
lugubrina, Glypta 17
luteofasciatus, Banchus 24

macropyga, Glypta 17
macrura, Glypta 18
maculata, Lissonota 21, 95, **99**
maculatorius, Syzeuctus 18
magdalenae, Lissonota 20, 69, **72**
magma, Lissonota 19
mammilator, Lissonota 19
maurus, Exetastes 23, 116, **118**, 149
medianus, Exetastes 22
melania, Lissonota 20
melanopus, Exetastes 23
Meniscus 19, 58, **60**
mensurator, Glypta 17
mensurator, Glypta 17, 49, **50**
meridionalis, Lissonota 21
microcera, Glypta 17, **49**
minor, Exetastes 23
monileatus, Banchus 24
moniliatus, Banchus 24

monoceros, Glypta 16, 33, **36**
monstrosa, Glypta 18
moppiti, Banchus 24, 123, **125**
murina, Alloplasta 18
mutator, Lissonota 21, 84, **87**
mutillatus, Banchus 24

nigra, Lissonota 19
nigrescens, Lissonota 19
nigricornis, Glypta 17, 43, **46**
nigricoxa, Glypta 16
nigricoxa, Lissonota 20
nigricoxis, Lissonota 20
nigridens, Lissonota 21, 83, 95, **100**
nigrina, Glypta 17, **40**
nigripes, Exetastes 23, 117, **120**
nigriventris, Exetastes 23
nigriventris, Glypta 16
nigrobasalis, Lissonota 20
nigromarginatus, Banchus 24
nigrotrochanterator, Glypta 17, 49, **50**
nitida, Lissonota 19, 60, **63**
nitida, Lissonota 22
nobilitator, Banchus 24
notatorius, Banchus 23

obscurata, Glypta 17
obscurator, Exetastes 23
obsoleta, Lissonota 21, 85, **89**
osculariatus, Exetastes 23
oudemansi, Lissonota 21

palaeanae, Diblastomorpha 16
palpalis, Banchus 24, 121, **124**
palpalis, Lissonota 21, 95, **100**
palpator, Lissonota 21, 96, **101**
paludosa, Glypta 17, 33, 34, **35**
papyri, Glypta 17
parallela, Lissonota 20
parasitellae, Lissonota 21
parvicaudata, Glypta 17, 51, **52**
parvicornuta, Glypta 17
pedata, Glypta 17, 51, **53**
pellucida, Glypta 17
piceator, Alloplasta 18, **57**, 145
picticoxis, Lissonota 21, **91**
pictipes, Glypta 17, 51, **52**
pictus, Banchus 24, 123, 124, **126**, 150
piffardi, Lissonota 19
pilosella, Arenetra 18, **56**, 145

pimplator, Lissonota 19, 62, **66**
plana, Lissonota 19, 60, **63**
plantaria, Alloplasta 18, **57**
pleuralis, Lissonota 21, **93**
praebellator, Lissonota 20
procera, Exetastes 22
propitus, Banchus 24
provincialis, Glypta 17
proxima, Lissonota 21, 80, **81**
puberulus, Exetastes 23
punctifrons, Glypta 17, **53**
punctiventrator, Lissonota 21, 98, **106**
punctiventris, Lissonota 21, 99, **107**
punctulatus, Exetastes 23
pungitor, Banchus 24
pusilla, Glypta 18, **51**
pusilla, Lissonota 21
pygmaea, Glypta 18

quadrilineata, Cryptopimpla 22, 110, **112**
quadrinotata, Lissonota 22, 95, **100**

removator, Lissonota 19
resinanae, Glypta 18, **41**
reticulator, Banchus 24
rhenana, Lissonota 19
rimator, Lissonota 20
rostrata, Diblastomorpha 16, **31**
rubicunda, Glypta 17
rufata, Glypta 18, 42, **44**
ruficeps, Glypta 17
ruficornis, Diblastomorpha 16
ruficoxis, Lissonota 20
rufifemur, Lissonota 20
rufipes, Glypta 17, 18
rufipes, Syzeuctus 18
rufithorax, Lissonota 22
rufoclypeata, Glypta 18
rusticator, Lissonota 20
Rynchobanchus 24, 114, **126**

sabulosa, Lissonota 20, 68, **70**
sachalinensis, Lissonata 19
sanguinator, Banchus 24
saturator, Lissonota 22, 85, **90**
scalaris, Glypta 18
schneideri, Glypta 18
sculpturata, Glypta 18, **47**
scutellaris, Glypta 18, 40, **48**
segmentator, Lissonota 20
segmentellator, Lissonota 20

segrex, Glypta 17
semirufa, Lissonota 22, 95, **99**
serena, Lissonota 22, 98, **105**
sesiae, Lissonota 19
setosa, Glypta 17
setosa, Lissonota 19, 61, **64**, 146
sibiricus, Banchus 23
signata, Lissonota 20
signator, Lissonota 19
silvatica, Lissonota 22
similis, Exetastes 23
similis, Glypta 18, 43, **46**
simulator, Lissonota 22, 91, 99, **103**
sinuatorius, Exetastes 23
spectacabilis, Lissonota 19
stigmator, Lissonota 22, 84, **86**
strigifrons, Lissonota 21
subaciculata, Lissonota 22, 80, **81**
sulcator, Lissonota 19
sulphurifera, Lissonota 20
summimontis, Glypta 18
superba, Apophua 16
Syzeuctus 18, 54, **55**
szépligetii, Diblastomorpha 16
sziladii, Lissonota 21

tarsator, Exetastes 23
Telutaea 16, **28**
tenerrima, Lissonota 22, 90, **91**, 96
tenuicornis, Glypta 18, **41**
teres, Glypta 18
thomsoni, Glypta 18
thomsoni, Lissonota 20
thomsonii, Glypta 18
tibialis, Exetastes 23, 116, **119**
transversa, Lissonota 21
triangularis, Glypta 17
tricolor, Banchus 23
tricoloria, Lissonota 21
tristis, Exetastes 22
trochanterata , Glypta 18, 43, **46**, 144
trochanterator, Lissonota 22, 96, **101**
tuberculata, Lissonota 19

ulbrichti, Glypta 18, 42, **44**
umbellatorum, Banchus 24
unicincta, Lissonota 20
unicornis, Lissonota 19

variabilis, Lissonota 22
varicornis, Lissonota 21

varicoxa, Glypta 17
variegator, Banchus 23
variipes, Alloplasta 18
variipes, Lissonota 21
varipes, Lissonota 21
venator, Banchus 24
verberans, Lissonota 20
vernalis, Lissonota 20
versicolor, Lissonota 22, **93**
virgata, Lissonota 22, 85, **88**
vocator, Lissonota 19
volutatorius, Banchus 24, 124, **125**

vulnerator, Glypta 18, 43, **45**, 49
vulneratrix, Glypta 18

woerzi, Glypta 17, 33, **36**

xanthognatha, Glypta 17

zagoriensis, Banchus 24
zangezurica, Glypta 17

Colour plates

From specimens in the Natural History Museum

Plate 1. Tribe: Glyptini
Telutaea brischkei Förster
(page 28)

Plate 2. Tribe: Glyptini
Apophua cicatricosa (Ratzeburg)
(page 30)

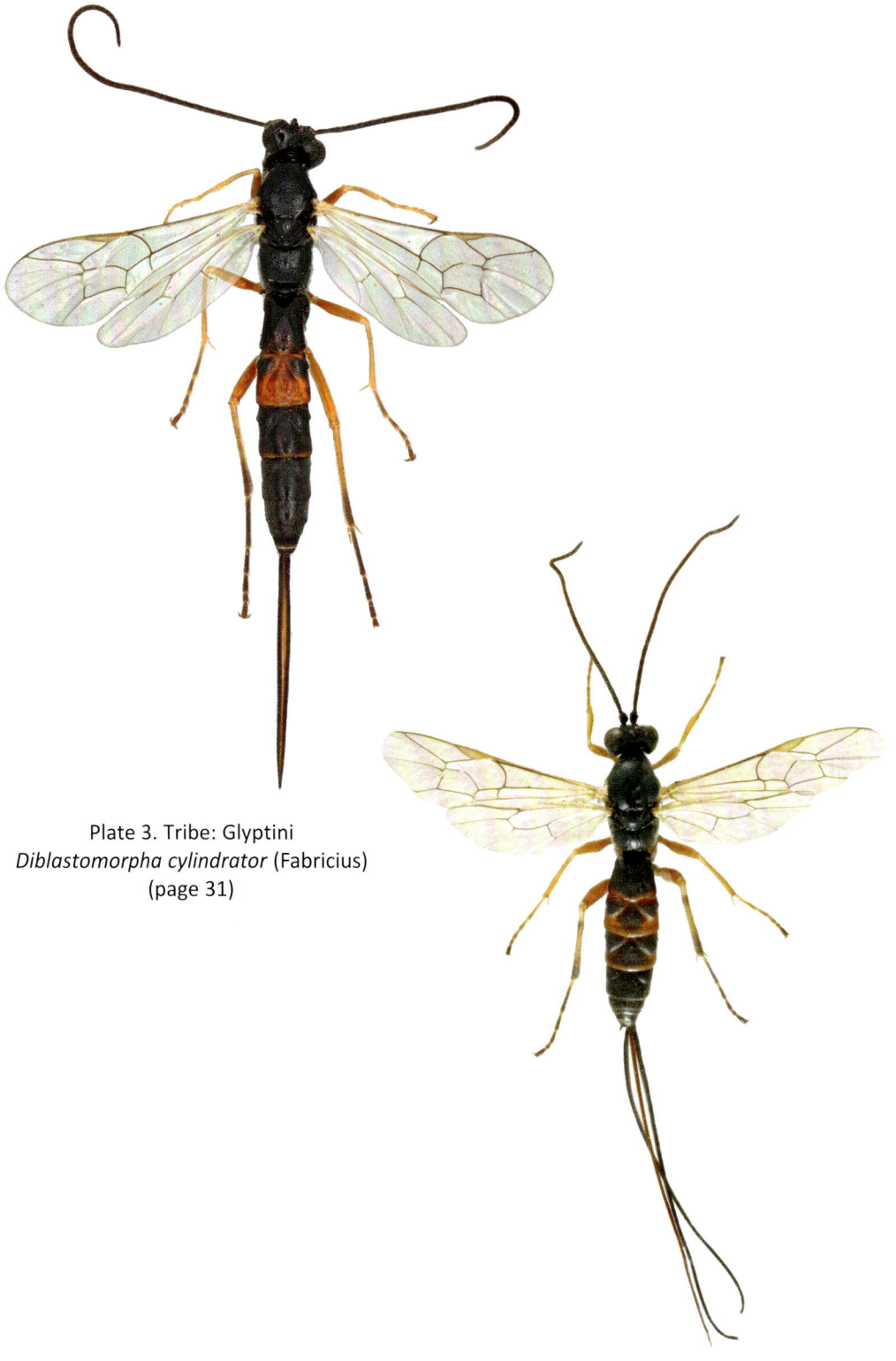

Plate 3. Tribe: Glyptini
Diblastomorpha cylindrator (Fabricius)
(page 31)

Plate 4. Tribe: Glyptini
Glypta trochanterata
(page 46)

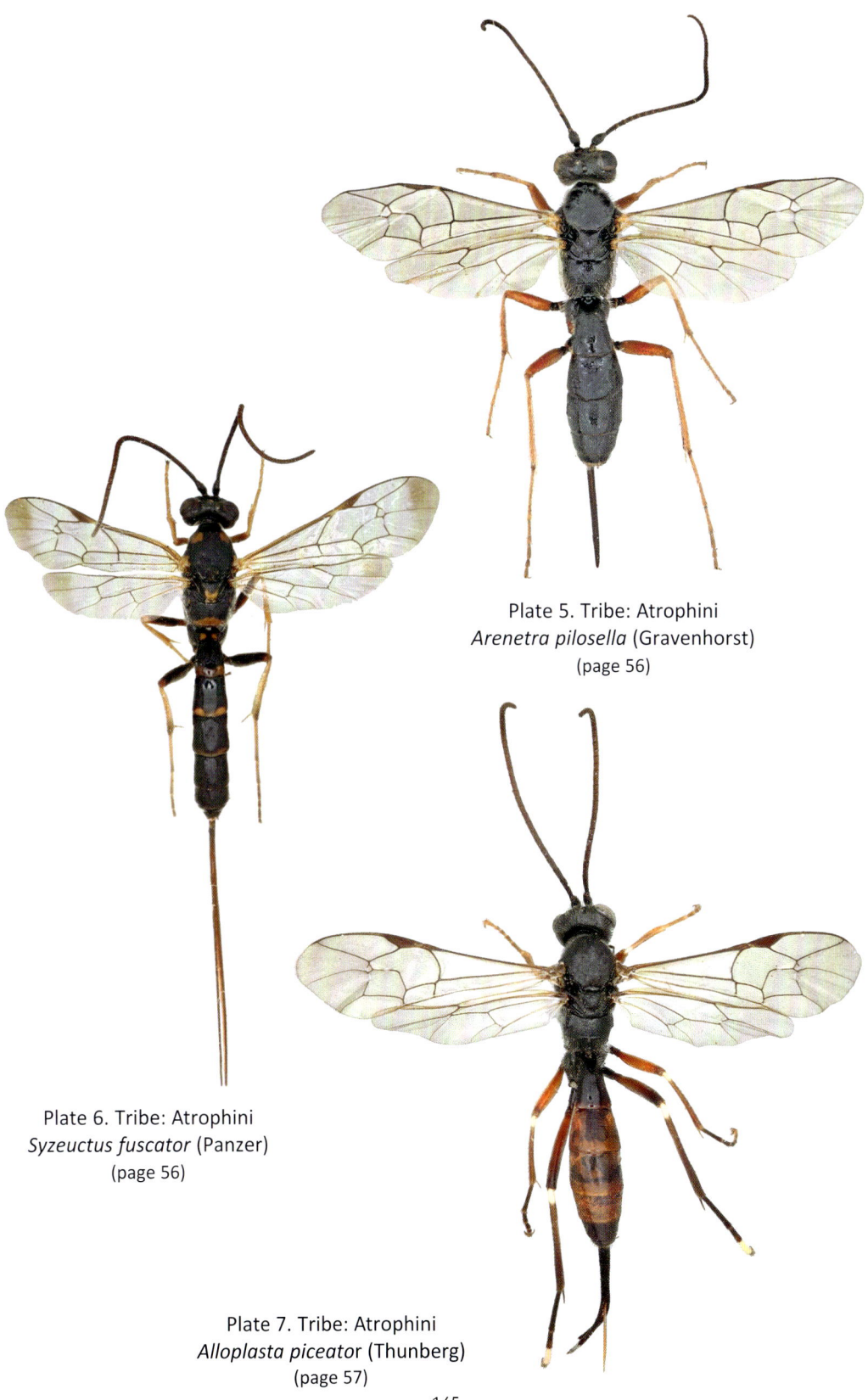

Plate 5. Tribe: Atrophini
Arenetra pilosella (Gravenhorst)
(page 56)

Plate 6. Tribe: Atrophini
Syzeuctus fuscator (Panzer)
(page 56)

Plate 7. Tribe: Atrophini
*Alloplasta piceato*r (Thunberg)
(page 57)

145

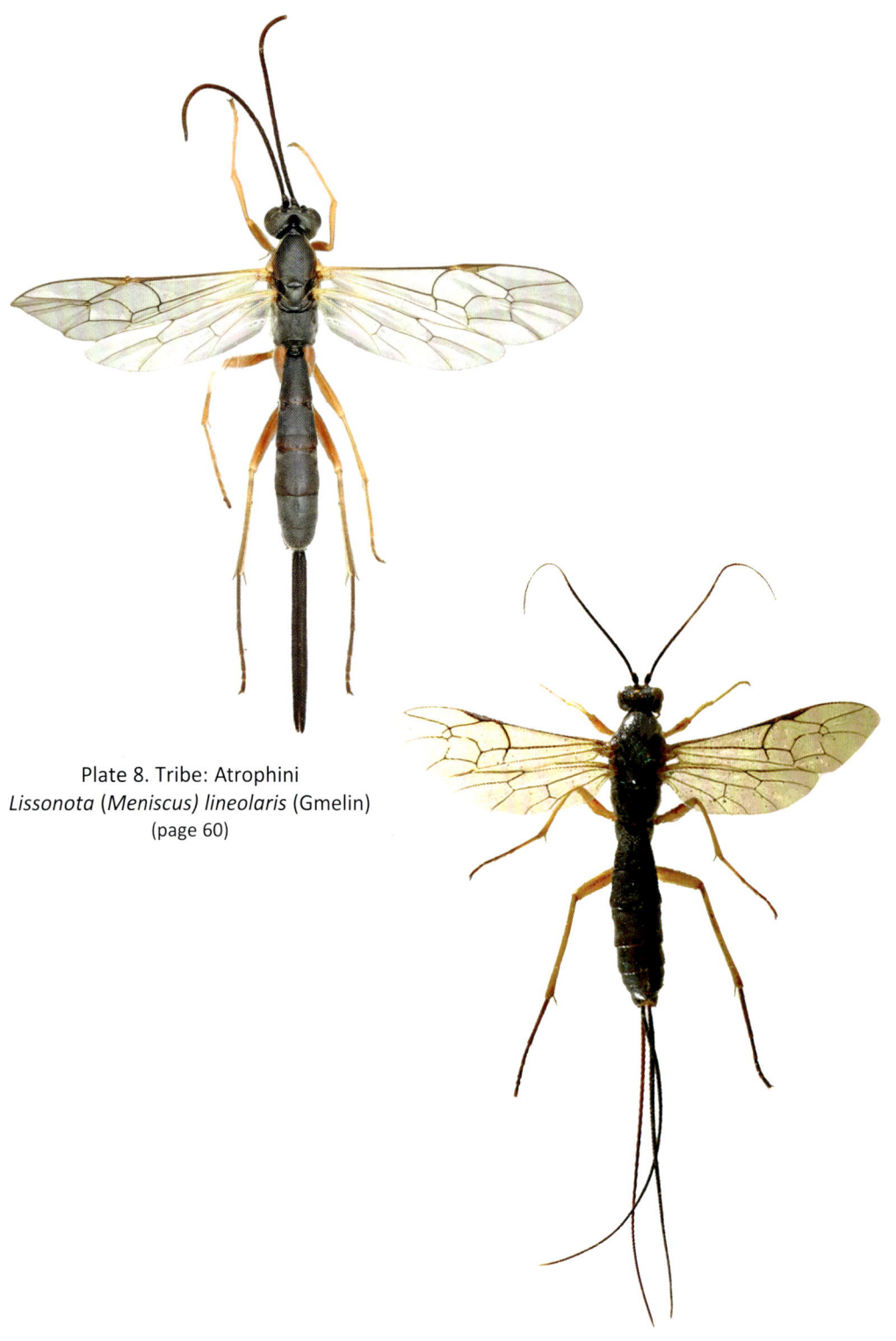

Plate 8. Tribe: Atrophini
Lissonota (*Meniscus) lineolaris* (Gmelin)
(page 60)

Plate 9. Tribe: Atrophini
Lissonota (*Lampronota*) *setosa* (Geoffroy)
(page 64)

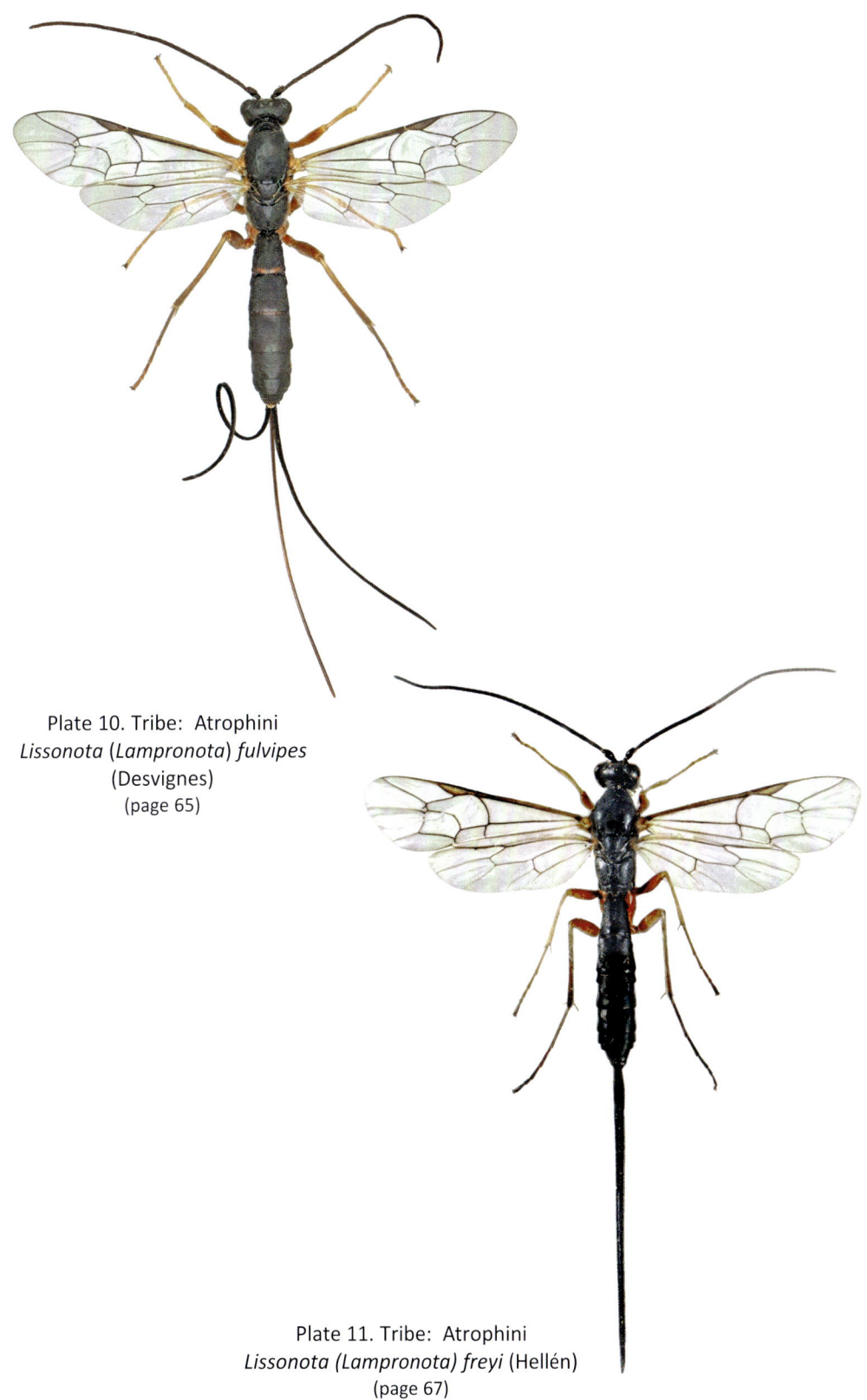

Plate 10. Tribe: Atrophini
Lissonota (*Lampronota*) *fulvipes*
(Desvignes)
(page 65)

Plate 11. Tribe: Atrophini
Lissonota (Lampronota) freyi (Hellén)
(page 67)

Plate 12. Tribe: Atrophini
Lissonota (*Lissonota*) *clypeator* (Gravenhorst)
(page 71)

Plate 13. Tribe: Atrophini
Lissonota (*Loxonota*) *histrio* (Fabricius)
(page 74)

Plate 14. Tribe: Atrophini
Cryptopimpla altipes (Holmgren)
(page 112)

Plate 15. Tribe: Banchini
Exetastes maurus Desvignes
(page 118)

Plate 16. Tribe: Banchini
Banchus pictus Fabricius [male]
(page 126)

Plate 17. Tribe: Banchini
Rynchobanchus flavopictus Kriechbaumer
(page 126)